T0331719

SUCCESSFUL MANUFACTURING TRANSFORMATION

RAPIDLY GET OUT OF MANUFACTURING
CRISIS AND ACHIEVE GLOBAL
MANUFACTURING COMPETITIVENESS

DR AZLAN NITHIA

PARTRIDGE

To order additional copies of this book, contact
Toll Free +65 3165 7531 (Singapore)
Toll Free +60 3 3099 4412 (Malaysia)
orders.singapore@partridgepublishing.com

www.partridgepublishing.com/singapore

To my late three children,
with love and everlasting memories:

- **Azwadi Aznil Nithia**
- **Azwari Aznal Nithia**
- **Aztika Azrin Nithia**

Contents

Preface

In the journey of writing this book, I benefitted from the experiences of many manufacturing experts and academics, both local and from abroad. I have brought together my vast years of experience by applying those learnings in high-volume manufacturing, crisis transformation, doing the right things—the right way, doing one thing at a time, engaging people, and operationalizing the *one-team, one-family, one-focus concepts*. I have a wide range of transformational experiences, implementing digitally connected operations, lean implementations, digital connectivity with automation and robotics, connected factory operations, and lessons from many experienced *senseis* (gurus *or* masters) from various industries around the world.

My early adoption of the lean principles and applying lean thinking in everything we did all began when I first read the book entitled *The Machine That Changed the World* by Jim Womack (he introduced the term *lean* to the world), followed by *Kaizen: The Key to Japan's Competitive Success* by Masaaki Imai (who acquaint Kaizen to the world), and *Toyota Production System: Beyond Large Scale Production* by Taiichi Ohno (who founded and introduced the TPS at Toyota). These appraisals started my relentless search for manufacturing excellence, and this search led me to numerous research into various manufacturing systems and concepts like Toyota Manufacturing System (TPS), lean practices, the importance of people engagement, creating a winning team, and the influence of a high-performing organisational culture in the competitive manufacturing operations.

Over the years, I have worked with many manufacturing organisations in different parts of the world, particularly those engaged in high-volume mass manufacturing operations. I have spent more than forty years championing various manufacturing functions, making major transformations possible, introducing digital product development, automating high-volume production, digitally connecting smart manufacturing operations, and successfully transforming manufacturing organisations out of crisis to achieve global competitiveness in productivity and cost.

This book will give you the practical guidance to multiply your current performances and resources, to achieve more with better results together with people's commitment to go beyond and to *win* like champions!

This is an excellent practical guidebook for every leader and manager in a manufacturing operation or organisation. For those in difficulty to achieve results or cannot deliver the results, facing a near closure, or in a crisis or organizations that want to achieve global competitiveness.

I wish you great success in your journey to achieve manufacturing excellence!

Acknowledgements

Welcome to my third book!

My first book was published in 2018, titled _Transitioning into New Manufacturing Paradigm._

The second book was published in 2019, titled _Achieve Manufacturing Excellence Lean and Smart Manufacturing._ In this book, I further expanded the important concepts of customer focus and implementing lean manufacturing as a foundation to achieve smart manufacturing. This is important as the critical foundation required for the readiness and implementation of smart manufacturing as well as pursuing the implementation of digital technologies, the internet of things (IoT) and to achieve the status of the _factory of the future_, also called a _connected factory._

This is my third book, it is a practical handbook for any leader who wants to get the manufacturing organization out of a manufacturing crisis successfully. This book is equally an important handbook to improve the current manufacturing operations so that it does not get into a crisis situation and to achieve competitiveness globally. For leaders who intend to prevent their organization from ever getting into any kind of manufacturing crisis mode, this book will be a good guidance and it is included with all the required tools to achieve it. No leaders intend to be confronted with any kind of manufacturing crisis or get eliminated. This book is for every manufacturing leader in any type of industry and for those leaders who want to become competitive and successfully compete globally.

This book is written by the author based on his vast years of experience in successfully transforming organizations out of extreme manufacturing crises (of near closures) and advancing these organisations to achieve global competitiveness at multiple sites, multiple countries, and different situations, managing different cultures and championing the safety and quality behaviours in everything that was done by the people.

This book will serve as an important *one-stop* reference handbook that offers various practical solutions that were proven at several sites, a step-by-step guidebook to any leaders who want to create and achieve enduring and sustainable manufacturing.

It is based on proven successful manufacturing transformations on how to get a manufacturing organization out of an extreme operational crisis (or nearing a close-down or shutdown), transforming an organisation in crisis and turning it around from a manufacturing crisis and finally being able to achieve global competitiveness.

This book was written based on extensive work experiences, learning from failures and successes, experimentations, and proven approaches by the author himself. It references the author's various manufacturing transformations achieved at large-scale manufacturing organisations and how the author re-positioned those organisations into a competitive operation. These are all based on his hands-on transformational leadership work.

I had the privilege to work with a person who had worked directly with Mr Maasaki Imai, and he wrote his book titled, *Chronicles of a Quality Detective*, which profiled the late Dr Shrinivas Gondhalekar (aka Dr G). Dr G introduced to the world one of the most powerful approaches towards solving quality problems by using a simple methodology called *differential diagnosis* (DD). I had personally worked with Dr G for almost twenty years, solving hundreds of operational, quality, and manufacturing problems.

Thank you to Maj Dr J Prebagaran for his kind contribution to this book as written in Appendix 1: Leadership in Action—Facilitating Learning Transfer for Performance Improvement.

I would like to thank the late professor Dr Gondhalekar for his kindness in contributing his amazing expertise in solving the un-solvable problem case study as written in Appendix 2: Problem Solving Case Study—a Case of the Porous Castings.

Introduction—Compete or Get Eliminated

We are operating our businesses in a borderless economy, and the world is the marketplace. We compete with organisations around the globe, and it is extremely important that we are always thinking about global competitiveness versus thinking locally or within the country. Even though we operate our businesses locally, we must always be driving towards <u>global competitiveness</u> in productivity and cost. We must be able to manage high complexities (low volume with high mix), constantly reduce lead time, and deliver good quality and competitive cost. Always think globally while operating locally.

If we are producing a product, out of Malaysia for $1 each product (as an example) and if the same product can be produced in Vietnam at $0.70 each, that is 30% lower than Malaysia and this translates to a 30% less cost benefit for the customer who is buying this product. It is an easy cost decision, assuming the quality and lead-time is similar. It will be a matter of time for the Malaysian producer to start losing their orders if they are not able to compete with Vietnam at the same cost or preferably at a lower cost. The product from Malaysia could easily be resourced or transferred to the lower-cost operations in Vietnam or to some other country that can offer a better cost and lead time. This is how we are operating in a borderless economy, and we are required to be constantly benchmarking, improving, and competing globally.

In today's world, we are operating our businesses in the fast-moving and rapidly changing economy. It is all about speed to the market (how quickly we are able to deliver our products to our customers). I believe in today's business competition, it is not about the 'big fish that eats the small fish', it is the big companies that are taking over the small company's businesses), instead it is the 'fast fish that eats the slow fish'; it is the companies that are able to move faster and will capture the businesses of the large companies that are moving at a slower speed in the market. For example, new start-up companies can rapidly capture their targeted market. They are small, but they are very agile, flexible, and able to get their products to the market much faster than the large companies at a competitive cost.

Getting our products to the markets or to the customers faster than our competitors is extremely crucial. (This is the power of winning in the borderless business.) It is about survival of the fittest and the *fast fish* winning in the marketplace. In a highly sophisticated supply chain that is driving businesses today with sophisticated digital technologies that are available to every organisation. It is important that companies adopt and adapt them very quickly to optimise all those available supply chain technologies effectively, optimising and applying them to compete or to optimise their resources globally.

If an organisation is always thinking about global competitiveness and constantly driving its organisation towards world-class efficiencies, delivering competitiveness in cost and the best lead times, then it is almost impossible to get eliminated by any other competition.

An organisation will need an experienced, proven, capable and visionary leader, who is empowered and entrusted to drive transformations and to achieve competitiveness. These leaders will have to be constantly focused on increasing operational efficiencies, reducing lead time, constantly finding opportunities to reduce product cost, constantly improving quality, and improving productivity in every function and with efficient inventory management. It is all about staying ahead of the competition in every competitive metric or KPIs.

It is not about being the number 1 in a particular business, but it is about being the most efficient and profitable in the business that we are operating in. The ability to sustain global competitiveness is always much more important and difficult than trying to be just the number 1 company. Any organisation could attempt to be number 1 in sales dollars, but deep down, these organisations could be inefficient organisations, and ultimately these so-called number 1 organisations can go down the profitability tube quickly, and even face closure. Therefore, focusing on being the most efficient and most profitable and being number 1 in the industry is meaningless if the organisation is not efficient and if it is not profitable.

If an organisation is constantly thinking about becoming the most efficient business, being a competitive manufacturer and constantly improving to achieve an efficient business status globally, then it is almost impossible for these organisations to ever get into any kind of manufacturing crisis or face any potential business closures.

Making small incremental improvements constantly is much more powerful than trying to sustain performance . . . organisations must <u>go beyond the mindset of sustaining</u>.

The manufacturing organisations that are slowly starting to slip away from world-class efficiencies and indications of their KPI metrics start losing their global competitiveness in cost, lead time, and productivity and will not be able to compete with their competitors. These are the organisations that will easily get into a manufacturing crisis, losing their competitiveness and finally losing their business. These companies will ultimately get wiped out of the business.

It is also important to recognise that the manufacturing sector has gone through a major evolution since the First Industrial Revolution through the years up to the current digital era of the Industrial Digital Revolution that has introduced high levels of digital connectivity and sophisticated automated human-machine interfaces. The manufacturing industry is constantly evolving from the use of intensive labour to one that uses smart automation and robotics to increase production efficiency, to increase agility, to reduce the cost of products, and to be able to respond faster to the customers' changing needs.

Advanced digital technological applications require a strong lean manufacturing foundation that enables the readiness of the organisation to implement smart manufacturing technologies or the factory of the future (FOF). It requires predictable advanced technological processes and well-trained people. Skipping the important phase of creating a strong lean foundation and an embedded lean culture within the organisation can be very damaging to the organisation's future, especially if these foundations are not implemented successfully.

Example of an efficient automation

Automating an individual process may not be the answer to reducing cost and to increasing operational flexibility, but it is about improving the whole system's efficiency. An automation initiative must increase the process productivity and enable the ability to run multiple different products (varieties) than before implementing the automation, much faster model changeover lead time than before, better process quality, and safer operation for our people. It must clearly position the organisation to become a globally competitive operation. If automation does not achieve this important advantage for your organisation, then why bother automating? We must find a lean solution first; we must be leaning our processes into automation and not automate first rather than later try ways to lean it.

Do not get trapped in 'automating waste'.

An efficient manufacturing organisation must pave the way to enhance the company's customer responsiveness and better productivity, constantly reduce lead time, reduce labour dependency, reduce product cost, and constantly improve quality. It is how we achieve all these benefits without increasing costs.

To prevent an organisation from getting into a crisis or non-competitiveness, it is important for every manufacturing organization to embark on lean and smart manufacturing so that these organisations will never get into a manufacturing crisis mode. I strongly believe the transition into lean and smart manufacturing paradigms is important for the survival of any manufacturing organisation. If an organisation wants to achieve manufacturing excellence and compete globally, it is important for the organisation to have a customer-centric or customer-first strategy.

Ultimately all industries will remain as business entities but will have transmuted into smart connected entities in a digitally connected business world. These smart businesses will have the innovative twist of automation and transformational digital technologies of business models and processes that increase profitability. The decrease in product costs will enhance customer delight, optimise consumer loyalty through lifetime value, and increase global market share with innovative growth while still remaining relevant and responsive to any market digital disruptions.

Building the kind of management and organisational culture in which everyone can contribute directly to adding value for the customers is important, keeping the lean and Kaizen spirit active to constantly improve and strengthen the manufacturing foundation so that smart manufacturing can be efficiently implemented and able to deliver the required results and customer responsiveness. To become a customer-driven company, the organisation must become a solutions provider, able to close the gaps in the supply chain and discern customers' needs that no one else is providing. The goal of value innovation must be the creation of top-line growth, especially sales growth to gain an increased market share in the industry that your organisation is in.

The success of an organisation cannot be achieved by force or by demanding performance. It can only be achieved by engaging everyone in the organisation, everyone coming together as *one team*, supporting each other, knowing each other's difficulties, understanding what success means for each other, and all connecting to *one mission*. It is always about winning in a *team sport* like true champions.

When a manufacturing organisation has gotten into a crisis, it is important to know how quickly the turnaround can be completed for an underperforming manufacturing company. What we expect is not to go wrong with an organisation. Sadly it can most likely go wrong. Even very well-managed corporations can have manufacturing organisations or sites that can easily get derailed into a crisis mode. Organizations that get complacent by showing good profits and maintaining good sales numbers can be the next organisations that may get into a crisis by losing their competitive edge.

An organisation's good profits and good sales results may not be a good indication of efficient operations. The organisation's performance must be based on the performance matrix of being better than before, this month better than last month, this year better than last year, and seeing the product cost going lower. Hours to make the product in production are constantly being reduced and the lead time is constantly being reduced. If these improvements are not measured, regularly and constantly tightened, this is another indication of the organisation getting into a complacent mode. Profits and good manufacturing efficiencies can be achieved by allocating higher than required production hours (loose standards) to the manufacturing operations, and the operations may show high efficiencies. However, that performance is against an inflated operational standard. Always seek to establish global standards, bridge the gaps, and constantly improve the standards further.

How will a leader know that the organization's performance is sliding towards a crisis? What are the early tell signs?

We as leaders must read the signs of performance sliding backwards if we are not tightening the operational standards (but instead loosening the standards), missing our production targets, and missing our delivery commitments to our customers; have rising costs; overhead expenses keep increasing; manufacturing efficiencies not reaching the optimum performance; have the quality of the products sliding down while increasing reworks and rejects; and have increased safety incidents and profits and losses exceeding budgets. There are many indications or signs that will show us that as an organization we are sliding backwards.

It will not be that all the signs are happening to drop at the same time. Instead it will always start with only one or two signs or KPIs of missing the targets in one or two important metrics. Later it slowly starts to spread into many other key KPIs. This is when it comes too late, and the organization is already going down the tube in the performance into a possible crisis mode.

To stop or prevent an organization from getting into a full-blown crisis mode, the leaders must take quick actions as and when negative performance signs are showing up. The actions must be taken rapidly. Allowing any of the crucial performance indicators to slide down for too long is what gets an organization deep into a manufacturing crisis or even a business closure.

Remember: Bad news does *not* get better over time.

Organisations must allocate time to frequently review and evaluate the current metrics, continuously tighten the standards (using lean and Kaizen methodologies), assess its performance against global competitions regularly, and do it with functional leaders.

Capable and knowledgeable leaders will understand how the KPIs drive an organisation towards achieving global competitiveness.

If the organisation's performance metrics are not reflective of global competitive metrics, this is another sign of operational downfall. We cannot operate an organisation in isolation, we must understand that in today's borderless economy, and we must think about global competitiveness. We must operate locally but it is important to be thinking globally and driving performance towards globally competitive productivity. There are no other easy ways to achieve continuous business sustainability and survival. It is always about the survival of the fittest in the long run.

It is extremely crucial and important to appoint a very experienced, knowledgeable, and capable leader to manage an organization out of a crisis. Appoint a leader who will respond rapidly to changes and take full advantage of the organisation's people culture and capability!

Organisations must appoint a very capable leader to lead, who is operationally very knowledgeable, proven in other crisis turnaround assignments, and with an attitude of never being satisfied or never complacent with current performance and able to drive continuous improvements in all the KPIs.

Organisations, especially those in crisis will need a stronger leader who

- the senior leadership team trusts and respects;
- can keep things very simple but very focused (doing one thing at a time);
- takes small steps (but with a clear strategy);
- will always be with the people in the Gemba (where the action is) constantly working together instead of telling. Walk the talk;
- will be a convincing communicator to engage the minds and hearts of everyone;
- is able to create high levels of confidence within the leadership team; and
- Is able to emotionally connect everyone as *one team, has one goal, and has one culture* throughout the whole organisation (leadership team and down to cleaners).

We need a have-experienced-it-all leader, an experienced and proven to have done it before at other sites, a leader who can put the organisation in crisis above his or her personal priorities 'with boots on the ground' and be daily with the people in the shop floor helping the people overcoming every challenge rapidly.

One important job of the leader in an organisation in crisis is to create an organisation that will transform to connect the people's minds and hearts to perform their tasks with full commitment, dedication, and passion, working tirelessly, working hard, and always going beyond the limits of time and relentlessly performing the task till it is completion.

Figure

CHAPTER 1

Managing out of Crisis

Leadership Challenges

When an organisation is in a deep crisis, it is very important to acknowledge that the employees' morale in the organisation is already at a very low level. Everyone would be working in their own functional work silos, and the people are not sure what the future holds for them. The team may already have lost hope in the organisation and in the work they do, and they are aware of being branded as a losing team or an organization awaiting closure.

There will be many issues and problems that will exist in an organisation that is faced with this crisis or near closure.

<u>**Below are some of the most common indicators of organisations that are in crisis or nearing closure:**</u>

1. Lack of customer focus—frequently missing customer orders, not meeting the on-time and full commitments to the customers, has many customer complaints, order cuts, and reductions.
2. Missing shipments and commitments—low adherence to daily tasks, constantly missing the weekly and monthly production adherence to quality and models.

3. Financial losses—monthly big financial losses due to missing sales targets, the increasing cost of production, deploying extra labour/resources, high cost of obsolescence, high inventory holding cost, missing payments, and unexplained losses.

4. Low productivity—low manufacturing efficiencies at processes, low assembly productivity, and not being able to meet the minimum required productivity performance.

5. High inventory and huge storage spaces—poorly managed plantwide inventory, missing inventories, inventories difficult to locate, frequent inventory adjustments, high inventory holdings in the warehouse, high obsolescence, inventories (parts for assembly) not available at production assembly lines on time and not in full.

6. Non-reliable equipment—machines not performing, frequently breaking down, equipment not maintained, poor equipment efficiencies/performances, and operating at very low overall equipment efficiencies (OEE).

7. Poor quality—low-quality performances, high rejects, lots of rework, high rejections within and in subsequent processes due to low in-process quality capabilities and lack of source quality.

8. Poor safety—high safety incident rates, frequent employee injuries and fire incidents, repeated occurrences of the same incidents, and employees not having a serious commitment to safety.

9. Poor audit compliance—compliance to labour standards, external audit failures, non-compliance to required local standards (labour requirements and union relations).

10. Poor union relations—union relations are poorly managed and low trust on both sides between the union leaders and the management team.

11. Missing KPIs—performance metrics not well defined, constantly missing on important performance indicators, repeatedly missing on the key performance indicators (especially those KPIs compared to other regional sites).

12. Poor teamwork—people working in silos lack of common goals, leadership team is working within their own functional

boundaries, and teamwork is not visible and lots of finger-pointing.

13. Poor people engagement—very low employee morale, low people engagement, low motivation to improve, and low commitment to deliver more and can be many more.

It is common to start pointing fingers at the people instead of focusing on the real issues and repeatedly telling the people how bad they are performing or how poor is their operations or their teamwork.

Is it the people or is it the leadership's failure? It does not matter now. The damage had already been done, and the organisation is now broken and is in a crisis.

There are many things to be fixed, but where do we start? Getting out of a crisis is not like completing a project, a goal, or a target. It is a huge mission; it is a journey, not a project. Only with one option or outcome expected—succeed or closure ('win or die'). It must be a mission or a journey that must be very convincing to all the employees. This mission must give the important needed hope that this journey will be a big success. It can be achieved for sure, and *we all will win like true champions!*

Getting out of a crisis is a huge mission. It is a very difficult journey and requires lots of hard work. But how will we *win* together like the *true champions*?

Leading an organisation out of crisis requires a leader with strong self-control and is humble but is results- or outcome-focused This is an extremely important mission, but it is very difficult to manage (self-control) by the leaders in times of extreme challenges. Emotions will run very high in every functional leadership team. It is critical that we as a leader who will lead an organisation out of this crisis always remember not to allow our emotions to overpower our intelligence. Maintaining self-control and understanding and respecting all employees in the organisation are powerful leadership strengths of a leader that is required.

Tough times require a tough leader (a tough boss) who will work very hard, tirelessly, and joyfully; able to enjoy the challenges; celebrate every little achievement with the people (however small it can be); and show high levels of confidence.

Don't allow your emotions to overpower your intelligence. It is all about your personal self-control in every challenging situation.

Be respectful to everyone in the organisation at every level. Everyone has endured a difficult time being told to do many different things. It is a fact now that the organisation is currently in a difficult position with a possible closure.

Let your honesty and respect towards everyone define your leadership style, make this obvious, and make it visually seen and felt by all.

The will of the leader must be equal to the will of the people in the organisation. Everyone is on this ship (organisation) together. We will sink or sail together. Developing the will of the people in this mission or journey is important, and it is hard. But this is crucial to gain the will of the organisation, which is the will of the leader.

Engaging everyone in *one common mission*, one common goal, and one common objective that <u>everyone truly believes in</u> and believes in what you say as a leader. This is important for the organisation's survival and success.

The mission is something everyone must commit to emotionally, full-heartedly, and relentlessly to achieve and deliver, everyone, as ONE TEAM, all 'Going Beyond' passionately. The journey is very hard, but everyone must feel the sense of achieving and progressing—the feeling of winning!

This is about going beyond the concept of teamwork. It is about how we will connect all employees in the organisation emotionally with their hearts and minds.

There will be anchor draggers along the way. Get them on board, or remove them quickly.

In times of challenging manufacturing crisis, we need our people, all the leadership teams, and all the employees to believe in us as leaders that they would truly trust. They should believe in what we say and do and the challenging mission that we are taking together.

It is all about the *power of our people*, not the machines, the robots, the automation, or bringing in consultants. We as leaders must believe (truly believe) in our people. Don't ever fake your belief with your people, show honest respect for what we already know, and shine on their capabilities and the miracles they can achieve and win.

The Most Important First Step

The first step a leader takes is the most important step, the first impression created is the most important impression, the first contact made with the employees is the most important contact and the first communication with the employees is the most important communication.

I have managed through difficult and challenging manufacturing crises in different countries. The approaches are similar, but managing through the people, situation, and local cultures during challenging situations are very different in different countries.

My first step has always been to gain knowledge of the local cultures, what the local people value most in their own local cultures, how relationship is natured outside work, and how bonds are created amongst the employees, their families, locals, favorite food, and so on. The local cultures and what the people value the most in Thailand are very different if we compare it to China.

I have seen many individuals fail in their job assignments (very quickly) in their foreign country postings, especially in leadership positions. Leaders who think that they can change people's local culture is a big mistake. For example, in certain countries, raising your voice while having a conversation can be considered very disrespectful or rude. Whereas in other countries, talking with a high voice is normal and acceptable. Giving a friendly hug in some local cultures is considered warm and welcomed, but in certain local cultures, it is forbidden. Therefore, understanding the dos and don'ts in the local culture is extremely important, and we can use them positively in the transformation journey.

As a leader, I have to gain a good understanding of the local cultures and the people, what matters most for the locals, and the behaviours that are liked and respected amongst the employees. Then it is important for me to look for ways (the connection) to use those behaviours and local cultures to integrate and align them in our journey. This is how a true emotional connection can be developed with the local leadership team and with every employee. Remember it is always about our people, not me. The more we give, the more we get. For example, give respect and recognise the current capabilities of every individual. The more we do these with our employees, the more positive their behaviours of the employees will be, go beyond the norm to deliver amazing results. This is how the employees gain pride in doing what they are doing and deliver better results.

Employees must gain pride in doing what they do and feel the success of their team. Success must be created from within by the local team! The local team must own the success, not any external individuals.

Never ever try to impose your personal behaviours or your home culture onto others in a foreign place. Learn to think like how the locals think, and understand what they value and how they connect with each other at work and outside work.

It is strongly advisable to undergo cultural familiarisation training before taking on any foreign country assignments. This will reduce the risk of making any obvious cultural mistakes, getting offensive, or demonstrating disrespectful behaviours.

It is also dangerous to assume that the organisational behaviours in an organisation will be the same as compared to the other even though both organisations are in the same country or in the same state. Your own family behaviours can never be the same as your friend's behaviours even though both your families are living across the street. Every family may look similar, but they are not the same. The same applies to an organisation. The organisation's structure or business may look the same as the other, but remember, they are not the same. I have seen many leaders make this mistake and later get into big failures.

Therefore, the first and most important step is to deeply know the people that you will be leading and understand their local behaviours, the organisational behaviours, and the people's local cultures. Knowing the people that you will be leading is very crucial, especially for achieving the first step to success of your journey with your people.

The employees in an organisation that have suffered a crisis for a long time, in a manufacturing crisis, or nearing closure are already with very low self-confidence and low morale. They have lost their passion and do not trust the leadership any longer, and they are already feeling like losers.

Where do I start, how do I start, who do I need? This will determine the actions and how far I can go along with the *team*.

Tell the truth about the current organization's performance, what is expected or the minimum requirement versus current actual performance, how big or how serious is the problem, the people must know the truth. The leadership team must hear it directly from me, as their new leader, telling all the truth as I know it and not hiding any truth from them. They must feel the truthfulness coming from my words, my emotional expressions are honest, and my actions are real as their new leader.

Meet the Customer Commitment

The next truth that needs to be told with full honesty (it is related to the customer orders) is <u>what can and cannot be done with the production delivery commitments and the sales commitments to the customers</u>. This can only be done after we have fully understood the seriousness of the operational and manufacturing process issues. This will require several weeks of spending time on the shop floor (doing the Genchi *gembutsu*) and observing every little thing that is happening in the production, equipment performance, quality (process and parts), processes (reliability), 5S, safety behaviours, front-line supervision, and inventories and supply chain flow. Understand and gain deep knowledge of the operations, know what is going on, know the operations gaps (versus what is required), and know who is doing what in the organisation.

Customer commitments must be corrected based on what can be delivered. This is a high priority. While an organisation is in a serious crisis, the customer's delivery commitments must never suffer and get an alternative supply. Deliveries to the customers must be quickly reorganised. This will require the full support and understanding of the bosses and the corporate planning team. There is no other way. The only way is to tell the truth of the current state, <u>commit what can be delivered, and deliver what is committed</u>.

The organisation was already missing their production commitments for many months and had already lost the credibility of the production commitments. Therefore, it is important to stop missing production or customer commitments and be realistic about what can be delivered under the current circumstances of a very broken manufacturing organisation.

> **Customer *delivery commitment*—commit what you can deliver, and you must deliver what you have committed. You only have one opportunity to reduce the orders or the commitment. <u>Credibility</u> is extremely important!**

The right sizing of the delivery commitments on a short-term basis while fixing broken operations is critical. There are no magical solutions to meet customer demands. I take a step back as far as required so that what I commit is what I will deliver. I cannot miss my newly reduced production commitments. I will have one opportunity to reduce my orders and re-commit my new production delivery plan. It is extremely important that I have understood the current production capabilities. As an example, I may have to reduce my production orders by almost 50% for the first six months and increase it for the following six months. After twelve months, I should be able to confidently commit to an acceptable higher production delivery plan. This may not be the most optimised production plan as it takes time to get an organisation out of crisis to deliver predictable, optimised production output.

The reduction in production and sales will have a big financial impact on the organisation in the short term. This must be understood by the corporate finance team.

The corporation must have a strong belief in your leadership capability that you will be able to turn around this organisation. It is in crisis now, and after we are out of this crisis, good delivery and financial commitments will be achieved. This belief and trust in your leadership must be strong and seem real at the corporate level.

Customer delivery commitments cannot be changed several times (once is already bad enough), but I have only one opportunity to commit it right. Reduce the production order and make an honest possible can-deliver commitment to the marketing team. Remember—credibility is extremely important. If you lose it once, you have lost it forever.

The Important People Steps:

1. Understand the people and the local culture. Every Saturday, I will try my best to have a lunch with a selected small group of employees to know each of them personally and their families and for the employees to know me and my family better and for me to know them and their family. This is a very good session

to know about the people, what is in their mind, and their local cultures. I am also invited to the employee's house for dinners, and I started to understand the families better as well. Getting connected with the people and the culture is critical to success.

2. Be respectful to everyone in the organisation, and never judge anyone. Recognise the hardship the employees, what they have been undergoing and how they lost hope in the company.

3. Tell the truth (from your heart) about the current situation or performance and explain the seriousness of poor performance. Never blame anyone or finger point at anyone, and never ever blame the previous leaders.

4. I must be honest in everything that I say, and my actions must speak a lot about my honesty as their leader. The people must feel my honesty and truthfulness, and I must gain the full trust of them (the team).

5. Spend most of my time in the shop floor, observing the production, and be with the people and doing our Genchi *gembutsu* daily, including the weekends. People must see and feel our passion and commitment to this difficult journey.

6. Gain deep knowledge of the current situation of quality, safety, 5S, productivity, equipment performance, process capabilities, inventories, supply chain, and how the current manufacturing teams are interacting and managing the operations. It is not about judging what is right or wrong but understanding (deep understanding) of the current state.

Next Step—Doing Just the One Thing!

In times of manufacturing crisis, there are many things broken in the operations, many issues, and problems. There are also many problems and issues that are not visible yet but will surface as we go deeper and deeper into solving one problem or issue at a time. It is very common as I have observed. Most leaders want to solve many things at the same time, and they want to see results instantly. They want changes happening immediately, they want the results to be delivered rapidly, and they will have pages after pages of action items for the local team to act on.

Sadly this is a sure failure in the making, trying to show the boss or the organisation that we as the leaders have tons of actions going on. Leaders do not succeed this way.

The first thing is to *stop* doing the many things.
Focus only on the *one* thing, the most important
one thing for improving the performance.

I was told to take over the leadership of this organisation that was poorly managed for many years. Many past leaders have failed to deliver the results. It was the most important operation for the corporation because of its location. It was strategically located near their most important markets, and products could be delivered to their customers very quickly versus their other operations in Asia.

Several years ago, I read a book written by Gary Keller and Jay Papasan that was titled *The One Thing: The Surprisingly Simple Truth Behind Extraordinary Results* (2013). This is a book that clearly explains the importance of focus, the *one thing*. This book reaffirmed my thinking of the importance of being very highly focused by taking only one important step at a time or doing the one most important thing at a time.

When you are trying to catch two rabbits at the same
time, you will catch neither of the rabbits.

As was stated in the Gary Keller's book in page 7, the phrase 'Be like a postage stamp, stick to one thing until you get there' (reach the destination), this clearly states that one stamp on an envelope cannot or will not arrive (or be delivered) to two different addresses (or locations). When the stamp tries to do that, then the stamp (or you) will finally not arrive at either place. So be like a stamp, stick to the envelope till you arrive at the correct address (location), and stick to the one most important thing till you fully complete the task with the best results. Therefore, you must figure out the most important one thing that gives the biggest impact or the one thing that gives the biggest improvement.

I was given three years to turnaround this operation that is in extreme manufacturing crisis. It is nearing a closure as I was given in my previous assignments (about three years) to transform this organisation into a highly competitive manufacturing site for the corporation and become cost-competitive with other Asian operations. This was also the largest operation site in the world for the corporation. This operation site had never performed well as required and constantly missed their customer orders, had a high cost, had poor quality, had bad safety records, had very poor equipment conditions, had poor labour relations, and many more performance issues. Anything that could go wrong in a manufacturing site or operations was going wrong at this particular site.

When I first arrived at this site, there were almost eighty people from other locations of the corporation who were assigned to this organisation, all of them playing the consultant role and telling the local leadership team what they must and must not be doing. The local team was treated like a bunch of failed individuals by all those eighty individuals who were trying to run the company that was in an extreme manufacturing crisis. The local staff became the listen-and-follow team.

The first thing I wanted to do was to give the highest respect and confidence to the local leadership team and not to those eighty individuals who are playing the consultant-type roles. They all were talking and telling everyone in the organisation what to do as if they knew it all.

After spending two weeks at the site, I terminated all those eighty people who were coming in from another plant. The so-called experts were playing consultant's role contracts that were currently placed on this assignment to transform this operation. I sent all of them back because the local team needed to gain respect and credibility. The local team must work hard, and they must own the future successes.

This was a big shocking news for the corporation and to all those eighty people. They were shocked, and they thought this operation could not be operated without their leadership, guidance, and support. These so-called experts were treating the local leadership team like *kids* and instructed the local team to do as they were told. These eighty people, all had more than thirty actions each, and when I added up all their actions, it was hundreds of action items that were required to be completed by the local teams. Many meetings were conducted daily to review the status of those hundreds of action items.

Most importantly, the local leadership team got confused with my action of stopping all the eighty people. They were wondering what my next move or action would be, a new group of consultants' support or what will be my next step. There was a lot of curiosity.

I invited all my local leadership team (all managers and above) for a one-day meeting on a Friday away from the company and away from work at a nearby hotel. This was our first off-site meeting after a month I had taken leadership responsibility for this operation that was in severe crisis and with talks of the potential closure of this operation.

I felt very honoured because the corporation and the senior leaders had lots of confidence that I was the best person for the job. I was fully empowered to do what needed to be done. I have successfully turned around several near-closure operations into a sustainable cost-competitive operation, doubled and tripled the business volume and converted the operation into high-automation manufacturing sites.

We started the meeting by talking about our favourite food and local restaurants and decided where we should have our lunch. I am a food maniac, and the local food was truly super delicious. Then I briefly told the team what we would be discussing for the day.

I explained to the team the seriousness of the current state of the organisation.

I showed the team **_some simple charts of current performance and pictures_** *that were taken to show the current state of the operation, namely the*

- *poor quality, unreliable processes, huge rejects, and reworks*
- *numerous safety incidents, injuries, and fire incidents*
- *poor productivity and low manufacturing efficiencies*
- *customer orders—low adherence and many customer complaints*
- *poor conditions of machines and equipment*
- *poor state of the plant and facilities*
- *poor financial performance*
- *high product cost, not able to meet the targeted transfer sales*

The leadership team agreed with the current state, understood the seriousness and potential closure, and wanted to know my direction. Everyone loved their jobs and loved the company. They had the experience, were young and talented, and wanted to see the organisation succeed.

I told the team that I was happy that I terminated all the sixty-five members (so-called helping and consulting) who supposedly came to help turn around the organisation. There was a loud, thunderous clapping and shouts of joy from everyone. I had everyone talk about the reason for this joyful clapping and what it meant to all of them.

With honesty and confidence, I told everyone that we did not need an outsider or a person from outside the organisation to tell us what and how to do because the local team was very capable and experienced, had high energy, and knew how to turn around this crisis. This site would become one of the best in the world. I showed my honest belief and capability in everyone.

We called ourselves, in fact, we branded ourselves *the champions*. Champions are winners. We are all winners, and we will win. The confidence level, passion, and commitment from everyone were going sky high.

Deciding the Most Important One Thing

We cannot win or succeed by doing many things. We must win by doing only one thing at a time, taking the most important one step at a time. Choosing the most important *one thing* to do and do it really well is the winning formula. When you try to do several things at the same time, you will spread yourself very thin, and several things you try to do, all will suffer from the lack of focus. Remember there is only some much you can do and be highly focused at one time, the one most important thing.

When in crisis, it is common that organisational leaders and bosses want to see many actions, the general thinking will be the more the better, but the reverse is true, be very focused on one.

To win any championship, for example, if a person wants to win a badminton championship, it is extremely important that the person gets fully committed to the one badminton game and be very focused on the *one* thing that is badminton.

In times of crisis, what is important is *focus* and a very clear direction. Everyone in the organisation must clearly know and accept the focus and the direction.

Everyone must be very focused on the one most important thing with a clear direction, deliverables (results, KPIs) with baseline and targets.

Communicating and engaging all employees and all levels and ranks towards a common direction are critical.

> **We must choose the most important *one thing* and work on that one thing. Be very focused, go deep, and achieve amazing results on that one thing.**

I told my leadership team to print all those hundreds of action items they were told to do before I arrived on a piece of A4 paper. After they had done it and written it down, I told everyone to take all those A4 papers that were filled with many action items. I asked everyone to tear all of those A4 papers they were holding in their hands. Everyone tore the papers into tiny pieces, and then all was thrown into the waste bin that was prepared in the front of the meeting room.

The burden of doing the many things that were told by someone else to this leadership team is now in the trash. The people's anger and frustration (of how this team was treated) are now in the trash bin. The hearts and minds of everyone are now light, free, and ready to charge into a new course of actions which we all will believe in as one team.

The next step was to develop our own actions by the seven functional teams, which are important for the organisation to get out of the current manufacturing crisis. Every team developed the <u>one most important thing</u> (action), the one thing that every functional leader and their team must focus on. The one thing that will make the biggest difference to the company's performance.

The anger and frustration of how the team was feeling before was now thrown into the trash. The leadership team have now gained self-confidence, faith, and courage.

The seven team leaders and their members are now ready for new actions that they know are the best to get out of the crisis, and everyone will want to own it.

All the team members shared their one most important thing (that will have a domino impact on several other issues as well), its current baseline performance, progressive targets to achieve (results) in the next several weeks in the next ninety days, and how they will engage their own team members and engage with other functional teams to succeed in their one most important thing to show a significant improvement in the next ninety days.

Every functional department leader developed their one most important thing. Finally the total of seven most important things that were agreed upon are as follows:

1. safety improvement
2. quality improvement
3. production: efficiency and productivity
4. customer focus: deliver weekly Production as committed (on time in full)
5. inventory: readiness, accuracy, and management
6. overall equipment effectiveness (OEE) and machine downtime
7. people engagement

Each of the above seven team actions had about three sub-KPIs (key performance indicators) that show the current baseline and the target to be achieved in the next ninety (90) days.

The whole organisation will be focused only on that one thing per team, a total of seven things, and every ninety days the goals will be reviewed by the team. The team will tighten the targets. This is our new path. The journey of continuous improvement is deployed in the whole organisation. This is now our journey to get out of crisis—<u>ninety days and seven things to improve. This is our organisation's new mantra—</u>called the **90 + 7 challenge**<u>. The details of this 90 + 7 challenge will be explained later in the chapter.</u>

> **Every week, all the key staff join the 90 Days + 7 things weekly update, baseline and progress made versus the target (to be achieved in ninety days).**

Every week, the leadership functional heads together with their key staff would team up to present the weekly progress update and make improvements every week. This week was better than last week's performance. We <u>celebrated every success </u>(small and big) that was presented by every team.

This was a very important weekly update meeting. Every functional leader and I (as the general manager) would never miss this update, and being a minute late was not accepted. No one in the meeting was allowed to use handphones or laptops during the meeting. Each functional head with their team would focus only on *one thing*. We had seven teams focused only on one most important thing per team.

Communicate the Journey—Difficult but Possible

It is important for every employee in the organisation to know the journey—where we are now, where we are going, and what is our mission. The mission must be very rewarding and connect the hearts and the minds of the people. The mission must be very meaningful to everyone in the organisation so that all employees want to be part of this journey. The journey is very difficult but possible.

> **The phrase <u>difficult but possible</u> becomes an organisational tact line.**

The mission must always be divided into four steps, namely

1. the immediate-term actions,
2. the short-term actions,
3. the medium-term actions, and
4. the long-term actions.

It must be simple to understand and easy to connect with people, and everyone can relate to it easily.

I did not communicate the four-step transformation plan till after nine months when most of the critical and urgent issues were addressed effectively by the teams with good improvements in motion. I have seen many leaders in an organisation are very eager to communicate their strategy and their mission to all the employees almost immediately. The leaders must know what to tell, how to tell, and tell the people. Remember one step at a time. The same applies when we are communicating to our people, one thing at a time.

An important reminder will be that it is always not about us (leaders), but it is about them (our people). It is always about our people being deeply engaged in the immediate-term actions and able to envision or see the light at the end of the tunnel for the next step.

It is common for the newly appointed leader (of an organisation in crisis) to immediately organise an *off-site meeting* in a nice hotel for a couple of days to brainstorm to create a strategy chart and a mission statement with his or her leadership team, and then the following week, the communication team will roll down the new strategy and the new mission to all the employees. This is a big common mistake that new leaders make, especially in an organisation that is under an extreme manufacturing crisis.

When a person is drowning in the sea, what must we first do? Give the person the hope to survive and throw a safety rope to the person or the safety float so that he or she can be saved immediately. We do not ask the person if he or she wants a salary increase or a better insurance policy or if he is tired. But we must do what is right the right way and at the right time. The same applies to the leader who is appointed to save the organisation. What is the important immediate action, focus on this one thing.

Figure 1: The Four Steps to Transform and Organisation in Crisis

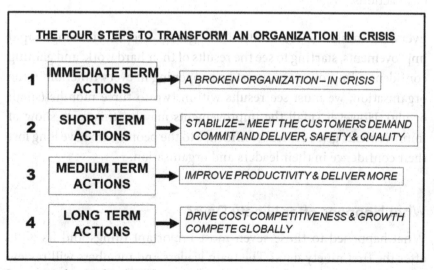

Source: Azlan Nithia (2023)

The people in a broken organisation that is in crisis and this manufacturing crisis need a clear, immediate direction and not a strategy to roll down.

The first step forward is an important direction for the people to gain immediate confidence and trust in their individual and their people capabilities to make the initial important changes or improvements happen rapidly.

We must know that the drowning person must be stopped from drowning, stop the person from drowning, and have his or her life immediately saved. We don't talk about the strategy or mission of tomorrow. We must engage in talks about the most important things to do now, the immediate first step.

People are not interested in your second, third, or fourth steps. People want to know the first step only, your step 1!

Similarly, with a manufacturing organisation in crisis, step 1 or the first thing is to save the organisation from closure by immediately taking action to get out of the crisis mode. Then later, when the team is starting to gain confidence, we can start discussing step 2—the short-term actions.

Everyone must be committed to going deep, achieving rapid improvements, starting to see the results of their hard work, and gaining confidence. Improvements must be happening in every function in the organisation, we must see results within two to three months (small or big). However small the improvements may be, this early show of improvements gives critical confidence to the people who have long lost their confidence in their leaders and organisation.

What Happens after Ninety Days?

What happened to those seven most important things and its KPIs after the first ninety days? The team leaders and members will review the baseline versus the current performance after ninety days. This was named as the phase 1.0 of the 90 + 7 challenge. It is now the time to celebrate the first ninety days' achievement by all the seven teams, it is an organisation-wide celebration. Everyone must know we have started to win a small win but an important first win.

After the ninety days, all the members of the seven teams got together and agreed on the new higher target (for all those same seven important things) to be achieved in the next ninety days. We called this phase 1.1 journey, and the earlier one was named phase 1.0 journey. This improvement challenge called the 90 + 7 challenge continued from one phase to another phase, and after every ninety days (or three months) the targets are tightened, better than the earlier phase, and the new targets are committed in the weekly performance tracking.

In a year (twelve months), the targets were tightened four times (or four phases) which will be phase 1.0, phase 1.1, phase 1.2, and phase 1.3. The commitment to make improvements at all levels became the new norm in the organisation as it engaged everyone in this 90 + 7 challenge.

This is a very powerful rapid-improvement program highly focused on doing one step at a time, and it is extremely important for the organisation to get out of the manufacturing crisis quickly. Everyone (including the operators and even the cleaners) in the company owned the success and clearly understood the mission. The direction of the organisation was well understood by all employees.

After six months, the leadership team and the employees started seeing some significantly big improvements in all those seven important focused areas. The confidence level started to increase, and the passion of wanting to do more was obvious in every team.

Team members came on weekends. They worked long hours, and there was a high level of passion to achieve and to show the weekly improvements committed in the 90 + 7 challenge. The culture of celebrating every achievement was practised in all areas, and employees got more committed to making improvements. More and more people wanted to be part of the 90 + 7 challenge improvement teams, and every employee now belonged to an improvement team.

Being part of the 90 + 7 challenge team was like being a member of an elite team in the company. Employees proudly wear special shirts and caps with special logos 90 + 7, and they were worn during the weekly update meeting.

The team also enjoyed a special lunch after this weekly update meeting. This lunch is part of the team bonding, bringing all team members closer as *one team* and *one family*. The energy level was getting higher, and the passion to win was very obvious. Everyone was pushing forward with relentless hard work.

The slogan was written on every employee's shirt—'We are the Champions'. This slogan was clearly displayed on the building—'The Home of the Champions'. Everyone could see this slogan as they entered the company. It is the place for the champions. Champions do not lose. Champions only win!

Attack the Problem, Put the Problem in the Center

Organisations that are confronted with an extreme manufacturing crisis will surely be confronted with tens and hundreds of problems in many parts of the organisation and processes that must be solved rapidly. Trying to solve many problems will also result in many challenges for the employees in the organisation because the tension levels among the people will be very high. This is when the leader's role becomes very delicate. They have to know the sensitivities and ensure the people are not attacking each other, not blaming each other but need an environment where the team is united to solve the numerous difficult problems and difficult situations by combining each other's intelligence.

Almost always, someone owns the problem, and the person who owns the problem will try to be defensive. He or she will try to give excuses or justify the reason for the problem. This is when the discussion about the problem gets ugly.

When an organisation is in crisis, there will be many problems that need to be solved very quickly. Arguing with each other about the problem will not solve the problem, but it only creates a bad relationship with each other.

There are two important things. One is respect for each other, and the other is the problem to be eliminated. The problem 'want to be a friend', a friend with the person who defends the problem. This causes an argument with each other because one person is attacking the problem, and the other person is defending the problem. This means there is a friend for the problem (the defender), and there is an enemy (the one who wants the eliminate the problem) for the problem.

Which side do you belong to? Are you the friend of the problem or the enemy of the problem? This sounds a bit funny, but this is always the issue. Which side do you belong to? Are you befriending the problem?

The problem will never get resolved or eliminated if we are the friends of the problem. If we are the friend of the problem, then we will attack the other person instead of jointly attacking the problem. The same problem continues to be reappearing all the time again and again.

The repeatedly reoccurring problems are the problems that were not eliminated previously. Remember no one likes to have problems, but problems do happen all the time, especially in an organisation in crisis. It will obviously have many problems to be eradicated. Blaming the other person for the problem does *not* solve the problem. It only creates happiness for the problem and not for both of you. This wastes a lot of time and resources of the team members blaming and arguing with each other. There is no time to be wasted, attack the problem together, and kill the problem.

Figure 2: Put the Problem in the Center

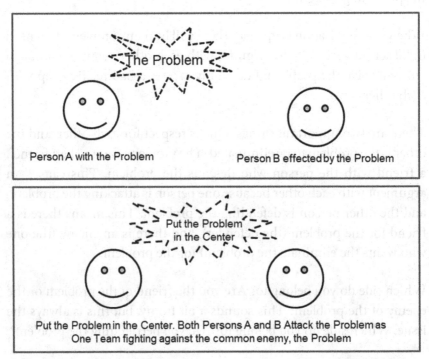

Source: Azlan Nithia, 2023

In figure 2, person A is responsible for resolving the problem because the problem had occurred in the process area of person A. Therefore, person A is the one with the problem, and person B is impacted because of the problem. Let's say that person A is responsible for producing a part of the moulding process and then delivering the part to person B who is responsible for the assembly process. The problem is that the defective part or component from the moulding process is now in the assembly process area. The assembly has to be stopped due to the rejected moulded part or component that cannot be used for the assembly process required for the product. As a result, person B misses his or her production schedule and cannot deliver the product as promised.

The argument will be person B is blaming person A for missing the schedule, and person A is trying to defend themselves (giving excuses of why the defect had happened). What should both of them do instead of arguing and blaming? This is when both of them (person A and B) must collaborate to put the problem in the centre and attack the problem (rejected part and why a rejected part flowed into assembly). Therefore, both persons A and B should be focused on the rejected part (the true enemy), discuss how to prevent this type of defect from occurring again in the moulding process, and if it ever happens again, know how to prevent that defective part from flowing into the assembly. In this way, both of them are looking for a solution to solve the problem, and they do not blame or fight with each other.

Be respectful to each other when faced with a problem. Attack the problem as your common enemy, never attack the other person.

The only way forward is to make the problem an enemy for both persons. This means pushing the problem to the centre, pushing it out away from the person with the problem. Now the problem becomes an enemy for both persons, the problem is not a friend any longer. Only the enemies are getting ready to attack. The problem gets to the centre only if we are willing to put the problem in the centre. This is the most difficult thing to do, but there is no other way. Now everyone tries to *kill* the problem in a collective manner.

Do not attack each other when you are confronted with a problem but be respectful to each other and go after the problem. Make it your common enemy to be *killed*. Kill the problem. It is easy to argue and fight with each other. Remember when you argue with each other, the 'problem becomes the winner'. Therefore, what you want to achieve—if the problem wins, we all are the losers and the company loses as well.

See the chapter on problem-solving to assist you in using a simple methodology in an effective way for problem-solving. This is the method I have always used successfully and taught many others.

CHAPTER 2

Manufacturing Strategy—
Connect with People

Strategy and Mission

Let us understand strategy briefly. Strategy is a set of actions or choices that will be executed by our people over a period of a few years (long-term choices or actions). It is important to remember that strategy is always about our people who will passionately commit to engage and execute all the required actions to deliver the expected results. Only our people will execute and deliver the results as they intended or planned. Many times, I have seen beautiful strategy posters on the walls of the meeting rooms and the aisles of the organisation, but the strategy is not for the walls but for the flawless execution by our people in the entire organisation by everyone.

> **It is critical to develop a strategy that will be understood the same way at every level in the organisation, including the production line operators and even the cleaners.**

The strategy is always about becoming globally competitive, and it is not about becoming the best company or the best in the world but to become competitive and compete globally.

26

The best is yet to be discovered. It all depends on how we want to compete and win in our industry. Strategy is fundamentally about the choices we make to deliver the required competitive value to a specific customer; this is the most fundamental.

There is always a better way and never the best way.

The strategy is embedded with the end in mind objectives and targets in a holistic manner (the whole organisation) over the long term that we want to be competing in and the unique position the organisation will achieve and would become (achieving superior performance and value).

Strategy must always have a high level of clarity and be easy to understand by all. It engages the people with absolute simplicity. Strategy must not be complex to understand, but it must be the simplest thing to understand by all (including the operators) in the organisation.

The strategy must keep everyone in the organisation hungry and very motivated to relentlessly improve and achieve higher targets to *win!*

Without a strategy, it is very difficult to connect the people, communicate what we are doing in the organisation, and show how it is connected to the transformation (out of crisis) with the actions or improvement programs.

Many times, we come across the organisation's leader creating strategies that are complex and complicated, and only managers can understand and explain.

We need a powerful connection and alignment between the strategy and the transformation journey. It is thinking about the whole thing (end to end)—the whole situation, the whole organisation, and the whole outcome to transform out of crisis successfully and ways to be able to compete globally. The strategy we developed in a manufacturing organisation that is in crisis (or nearing closure) will be unique. It may not satisfy everyone's needs (initially), but it must be focused on transforming and becoming competitive. As time passes, the results start to show up. Everyone gets the confidence, and they will passionately get involved.

First things first, the priority is to get out of the crisis and prevent a closure. The strategy must incorporate the whole thing but with the thinking of doing only one thing at a time and taking one step at a time. It is not a hundred meters race, but it is a marathon run that requires passion and endurance (physical and mental) from everyone in the organisation.

Next is how well and clearly we communicate the strategy to everyone in the organisation about where we were, where we are now, where we will be going next in the transformation journey, and how it benefits all and the future stability of the organisation. These are important responsibilities of the organisation's leadership team. We must make everyone live it and believe in the four-step strategy by executing the actions daily and connecting everything being done by everyone to achieve the results or the outcome. Make this week's performance better than the previous week. This month's performance should be better than last month's performance.

Remember to celebrate every little success with the *team*!
We work hard and play hard!

With many years of experience in transforming manufacturing organisations in crisis and getting out of crisis successfully, it is clearly proven that strategy, execution, and connecting our people must all go together. These must not be thought of separately but as one aligned and connected action. Strategy is all about getting it done and achieving the results. It is about execution through our people to get it done—by doing the right things, the right way, and at the right time, the three-rights concept.

Execution is about getting it done and making it happen. It is getting things done the right way by our people by engaging their hearts and minds with a strong belief and eagerness to get it done. Everyone must know this will lead us to the win, the ability to see the light at the end of the tunnel.

I have successfully developed and deployed a simple-to-understand strategy. It is easy for everyone in the organisation to get aligned. It is a one-pager and easy to communicate our strategy to our people. Most importantly it must make good logical sense to every employee.

Figure 3: Manufacturing Strategy

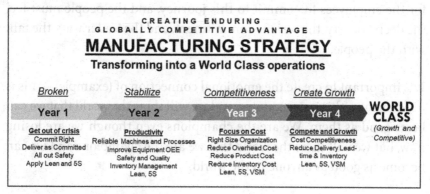

Source: Azlan Nithia (2023)

It is a four-phased strategy to get the manufacturing organisation out of an extremely serious crisis and reposition the organisation to achieve global competitiveness status. The number of years to complete an organisational transformation will be dependent on the seriousness of the crisis and the readiness of the people.

To effectively undertake a successful manufacturing transformation, it is important to understand that it is not a quick-fix mission. This must be well understood by the leader who is taking on this mission and his or her bosses who will be involved. It is a step-by-step transformation journey, involving people to successfully complete the transformation.

The concept of doing one thing at a time is a step-by-step approach, doing the right things in the right way and most importantly, doing those right things at the right time.

Phase 1 (Year 1) Broken Phase—Get out of Crisis

This is the most *fragile* moment for the whole organisation, particularly for the people, important to create an emotional connection to the crisis and the passion to drive the transformation. This process will require a huge amount of people's *trust*. Our people must trust the leadership team, especially the number one leader of the organisation. Everyone must feel this journey is safe for them and worthwhile for all the hard work they will put in, and most importantly, they should believe in this journey that 'We Will Win'. The leader must create an important reason for the employees to commit to this journey, and the people must feel the deep honesty, the serious commitment, and the leader walk the talk with the people.

It is important to create the emotional connection of (example)—this is my company, this is my country, and we will show the world that we can be as good as them. We are the champions even though we are losing now, but we will show the world that we are the champions and we can become as good as anyone in the world.

Our employees must believe and trust the organisation's leader full-heartedly, feel the openness of the leader's commitment, and all the leadership team members must be walking the talk. It requires a lot of heavy lifting to fix the many things that are broken in many parts of the operation. We need everyone in the organisation to come together as one united team and commit to fixing what is broken. The journey is exciting and everyone has a strong belief in the leader's strategy. Explain to all why it is extremely important to focus all our energy and efforts to get things done rapidly <u>but taking one step at a time and doing only one thing at a time</u>. The work culture of 'I commit and I will deliver' must be communicated. People must be convinced and be practised by everyone in the organisation.

Never overwhelm people by making them feel that everyone must be doing many things at the same because there are many things that are already broken currently. Instead, keep reminding everyone, to do only one thing at a time and must do it well—<u>do the right thing, do it the</u>

right way, and do it at the right time! This is an important mantra that I have always practised successfully in all my years of manufacturing experience.

The program, which was explained in chapter 1 called the '90 + 7 challenge' program (ninety days + seven improvement teams) must be started in phase 1. The seven teams will establish their current baseline performances and set the target to be achieved in the next ninety days. Each team is focused only on one key performance (or KPI). The performance progress is tracked and presented weekly by the teams to the leadership team. It is extremely important that the '90 + 7 challenge' teams are given the highest level of importance by the leader and by the leadership team. Show your commitment by attending the weekly updates by being there sharp on time. No one should come late for this important update meeting.

I have seen many leaders fail because they generally feel doing many things will solve the problems faster, but the opposite is true. Do one thing at a time!

As a leader, do not try to run the race without looking back often to make sure your people are running the race with you, together with you and running along with you. Look back often to ensure that you are not alone in the front without your team. Remember that your people must keep up with you on this journey.

Nobody may believe in us. Our organisation had been broken for too long (losing money for many years). Many people in the HQ may not trust what we say because this operation has failed to deliver on many commitments. This is when we must be strong emotionally and understand external sentiments. Whether outsiders believe in us or not is not at all important. What is important is that we all must believe in ourselves internally (our employees), trust each other in this difficult journey, and forget about what outsiders say or talk about us. We are *one team*, and together we will succeed in this difficult phase 1 journey of getting out of crisis. We *must* stay focused.

I have always used this phrase,
'It is very difficult, but it is possible!' Yes, we can do it!!

Doing one thing at a time and taking one step at a time increases the confidence level of the team to deliver full commitment, dedication, and relentless passion.

In every country that I was assigned to transform an organisation in crisis, the priority for me has always been to quickly do the following:

1. understand people's behaviours (due to the past) and their current thinking is important;
2. value and respect people's feelings and talk often with people to know their feelings;
3. know the local culture, talk to the people often about culture, and learn and understand that every country has different cultures;
4. know what everyone values the most in the organisation and go with the flow;
5. learn how respect is shown to each other by the employees and copy and practice the same;
6. know the words or phrases that are *toxic* that must not be used with the employees;
7. do not say words that will negatively impact the local people's feelings. What is right in another country may not be right in this country;
8. always be very respectful and show kindness to everyone;
9. frequently take small groups of people for lunch or dinner to know them better and for the people to know you better, especially knowing each other's thinking is crucial. Listen more and talk less; and
10. visit some of the employees' homes, eat with them, show your humbleness, and join some local events with the people. Enjoy some personal time with the people.

The above are ten important steps that a leader who is leading an organisation in crisis must quickly learn and practice. Do not impose your style or your opinions on others. A good leader must understand that in times of extreme manufacturing crises. The people's tensions and anxieties will run high due to their past failures, poor leadership behaviours, past humiliations, and the challenges in front of them currently.

By the end of phase 1, the teams (90 + 7 challenge) should be showing some good progress. I have normally seen the teams achieving no less than 25% or more improvements in all those seven areas of focus, improve 5S in all areas, lesser machine downtimes, lower rejects, meeting weekly production targets (the 'I commit and I deliver' culture), and processes becoming more reliable.

Phase 2 (Year 2) Stabilise the Processes—Strive for Better Safety, Quality, and Productivity

The stabilisation phase (phase 2) is equally challenging as phase 1. Everyone has tasted the difficult beginning of the transformation journey and experienced the feeling of achieving improvements and success through the important program called the '90 + 7 challenge'. The teams have seen their capabilities to deliver results in ninety days, reset a higher target for the next ninety days, and work harder to achieve it.

This is an important program to connect everyone, engage all, and continuously strive to better the performance. Every week should be a better week—the continuous improvement of work culture.

The 90 + 7 challenge program must include these critical areas. Below is an example of the <u>seven key areas</u>:

1. safety improvement
2. quality improvement
3. production: efficiency and productivity

4. customer focus: deliver weekly production as committed (on time in full)
5. inventory: readiness, accuracy, and management
6. overall equipment effectiveness (OEE) and machine downtime
7. people development and engagement

In all those seven key areas, each team focuses on one improvement. For example, team 1 is responsible for safety. This team will only focus on all areas related to safety, nothing less and nothing more, only safety. Team 1 would have established the initial baseline during phase 1. So the team will demonstrate in their weekly update, all the improvements being made and the progress the team have made towards meeting their goals or the KPIs. In phase 2, the 90 + 7 challenge program will start with phase 2. After ninety days, it becomes phase 2.1 and so on.

The teams must already be seeing some good results based on the improvements or progress they have made. The reference is the initial baseline in phase 1 and now starting with phase 2. Those improvements become the critical motivation for the team. The required confidence for the team to now start phase 2 and progress with better results every ninety days.

As the teams start phase 2, the organisation can already deliver better results in all those seven key areas (upon completing phase 1), and this translates into a better overall organisational performance as the teams start phase 2. The safety programs (which include the 5S improvements) are getting better each week. Similarly the quality is getting better each week and the same for all the rest of the seven key areas.

What type of leader are you? Are you a genius or are you a genius creator? We need leaders who can create passionate and high-performing leaders. Become a 'multiplier' (Elizabeth Wiseman, 2010).

It is important to constantly improve and sharpen managerial skills, leadership skills, supervisory skills, and problem-solving skills and do the kaizen and lean deployment in every area and motivational program.

Problem-solving, kaizen, and lean skills are very crucial during the time of getting out of crises and continuously improving performance.

Develop Capabilities in all Seven Key Areas

1. <u>Safety improvement</u>—start with basic 5S deployment in all areas. Train everyone in basic safety like holding the handrails when going up and down the stairs. Catch people doing unsafe acts, use unsafe examples for training, and educate everyone about unsafe acts so that unsafe acts can be stopped in the organisation. Assign a team of technicians to start improving all the unsafe conditions in the plant. There will be hundreds of unsafe conditions to be improved to get some external technical support as needed to solve the unsafe conditions. The team must share all their improvements in the weekly updates and show photos of before-and-after improvement conditions. Starting small group safety teams in every is really good for spreading safety behaviour throughout the organisation.

2. <u>Quality improvement</u>—there will lot of rejected parts and a lot of reworks, but this should not deter one step at a time. Focus on training the operators, line leaders, and technicians on the basic quality and the knowledge of at-source process quality. Make the processes predictable so that a source quality can be achieved. The employees running the processes must never allow defects to flow to the next operations. Stop the defect at the source and rectify the defects before sending it to the next process. The team updates weekly all their improvements and shows how the overall quality performance is improving.

3. **Production efficiency and productivity**—productivity is the outcome of predictable processes and good quality being produced at source. This translates into predictable good parts being delivered to the final assembly process. Good inventory control is equally important (by team 5—focusing on inventory improvements) to deliver the right parts with the right quantity and the time for the final assembly process. The line assembly stoppages must be measured and a five-minute waiting time

must be considered a stoppage and measured as a daily KPI. Every process must measure (by shift, daily, and weekly) its own process area efficiency (productivity), but the overall plant efficiency is measured by the final assembly process performance and meeting the daily schedule adherence (completing the right model and right quantities at the right time). The team shares the improvements in the weekly updates.

4. **Customer Focus: Deliver weekly production**—measure the missed orders and the missed shipments and measure the committed plan versus actual delivered, the 'on time in full' performance. All orders must be delivered to the customers in the right sequence as per the commitment. As teams 1, 2, 3, and 5 are improving continuously (this week's performance is better than last week's), these improvements will translate directly into improving customer delivery accuracy, delivering products 'on time in fill' (or called the OTIF). This team shares the improvements made, the KPIs, OTIF, shipment performance, and the assembly process delivery accuracies in the weekly update meeting.

5. **Inventory: Readiness, accuracy, and management**—this team's performance impacts every process area and most importantly, the final assembly. If the final assembly misses the schedule as committed, then the company (team 4) will miss the OTIF commitment with the customers. The assembly must get the right parts in the right quantity and at the right time. This is the responsibility of team 5, to deliver parts to the assembly at the right time. If not, the line stops and gets recorded as a line stoppage. Every five minutes is considered as one line stoppage. For example, a fifteen-minute line waiting for parts is considered as three line stoppages. Therefore, this team (team 5) must translate team 4's requirements accurately and ensure all process areas are delivering the right parts into the inventory locations timely. One important measurement for team 5 is the location accuracy of the parts kept in the parts warehouses. The purchased parts (parts coming from external sources) must also be delivered on time in full and placed in the

locations accurately. The measure of location accuracy of the parts must be frequently checked and audited to ensure close to 100% location accuracy (compared to the system's data). The material suppliers must never be searching for the parts. All the required parts must be available in the right quantity in the locations as shown in the system's data. This team will share their improvements in reducing the assembly line stoppages and location accuracy in parts and part delivery improvements in their weekly updates.

6. **Overall equipment effectiveness (OEE)**—this is the measurement of the machine's performance. The higher the OEE measurement, the better will be the machine productivity. The OEE measurement is the outcome of the machine output quality, machine availability, and machine performance. This team, team 6, has the task of reducing the machine stoppages due to various machine problems and poor conditions. Assign a group of trained technicians to continuously improve the machines and equipment in every process area. Develop internal technical skills (the maintenance technicians) and bring in experienced technical experts to train our internal maintenance technicians so that our own technicians can quickly identify machine or equipment problems and resolve them rapidly. The plant in crisis will normally have a lot of machinery that is poorly maintained. This requires good execution by the maintenance team and team 6. This team will measure the improvements in machine downtime, actions being taken to improve machine availability time (machine uptime), and OEE improvements in their weekly updates.

7. **People development and engagement**—focus on developing excellence in every part of the employee services like the cafeteria ambience, tasty and preferred food, good toilets, and good employee transportation services. Make the employees feel the emotional touch that the company truly care for them. The team of highly trained trainers should train all new employees, be their buddy till the new employees settle in, observe the new employee turnover rate, listen to their people, create schedules for

employees GM talks biweekly, and go to the floor and talk to the operators. The talent and training manager must be constantly looking for leadership and managerial and supervisory skills development programs (the right programs, at the right time) for the people. Listen to your employees' suggestions and to their complaints, show that you care for them by listening to them, and be approachable. This team 7 will present all the improvements in employee services, people development, timely hiring, and recognitions and measure the turnover rates weekly improvements in their weekly updates.

Continuously Sharpen the Leadership Skills

It will be good to take all the managers out of the plant for two days of training (strategically planned based on current needs) away from work for two days to sharpen their skills. I have found it very effective to do it every six months. Pick a good, impactful, relevant leadership program and continue to take these programs from one level to another higher leadership level. These programs are to constantly upskill our leader's leadership skills, and this upskilling must never stop. Do it every six months without fail.

Teamwork and understanding each other are equally important for all the team members. Take the teams (especially the leadership team) out for a team dinner function, including some games and have a fun time. This type of function will bring the team members closer, improve team bonding, and create better understanding amongst all members and the willingness to help each other. Go ahead to find some good reason to celebrate good performance every two months.

'Work hard and play hard'—you and your people.

As a leader, you should take the time or allocate time frequently to take some of your key people (about ten to fifteen people each time) for breakfast or lunch over the weekends (preferably on Saturdays). These kinds of small group events will give you a good opportunity (as a leader) to know your people better and to better understand their challenges.

For the people to know you better, create a better understanding of the mission and discuss ideas (theirs and yours, always ensure it is more of theirs) in a very relaxed setting (not too noisy environment).

You as the leader of this organisation in crisis must commit a lot of time, be with the people, and walk the shop floor, and you will have little time for yourself. This requires your personal sacrifice and commitment to successfully complete this difficult journey with the team. Make it visible!

As the various teams continuously deliver better results, the overall organisation performance has improved in developing our people, better leadership, having the right people doing the right jobs, improved skills and teams delivering better performance week after week. Expect to deliver at least twenty per cent (20%) improvements in phase 1 and another twenty per cent (20%) improvements in phase 2 or more improvements in all those seven key areas.

This kind of structured positive improvement gives the required confidence levels for the bosses and the leaders in the headquarters who are closely watching your leadership deliveries and the overall company performance improvement.

Phase 3 (Year 3) Competitiveness—Focus on Cost

In the competitiveness phase (which is phase 3), this will start focusing on increasing kaizen events that are mainly focused on cost savings, reducing cost in every financial line item in the overhead and labour, better material yield, sourcing material at competitive cost, and reducing inventory cost. The team (all the seven teams) may not have control of the material purchase prices as it will be mostly determined by the global market situation, but we can optimise the material usage in every product, reduce scraps, find lower cost options, and increase recycling activities to reduce scrap (for example, the rejected plastic parts that can be recycled or reprocessed to make new plastic parts must be intensified without scarifying the product quality).

The maturity and the experience of the '90 + 7 challenge' teams must have already reached a high level by now, and the teams must be very capable to continue driving improvements deeper into phase 3, just as the teams as proven in the various phases as in phase 1 and phase 2. The goals had been to continuously achieve improved targets every ninety days. Try visualising the amount of improvements the seven teams would have achieved after completing phase 2, obviously all the amazing results.

Looking forward in the beginning, during the start of phase 1, current results after completion of phase 2 would have seemed practically impossible, but now looking backwards to where we are now versus the original baseline, it looks unbelievable.

Phase 4 (Years 3–4) Create Cost Competitiveness

In phase 4, the team is ready to drive growth and deliver cost competitiveness simultaneously. Get the team engaged and committed to growing the business by an incremental 50% to double the business. Do not think small, for example, 10% or 20%. This is not the way to think of growing your business. Think of doubling the business.

The organisation will not be entrusted with more business and production if it is not able to prove higher growth capabilities and cost competitiveness globally. The team has shown their ability to successfully complete the difficult transformation journey. The next task for the team is to create the capabilities to grow the business and continue to reduce manufacturing costs (labour cost, overhead cost, and cost of materials).

The organisation must be ready to show the capabilities to do more and more with fewer resources. Remember what got us here is not sufficient to take the organisation to the growth path and to be in a competitive mode. We need the new formula to deliver the required new results.

Continue the high focus of phase 3 of the '90 + 7' program, going into phase 4. This program drives continuous improvements in all seven important metrics and key results in all the critical manufacturing metrics.

Now is the right time to start the next intensified focus program on financial metrics (labour and overhead) and to grow the production outputs (or sales). Create a '**cost competitive team**' (CC team) that will focus on key financial metrics.

Train the leadership team to understand the P&L expenses' cost drivers and how to take control of all those cost drivers. Assign leadership team members to each selected key cost metric (six to seven key cost drivers) and review cost improvements on a weekly basis. Challenge the CC team to show cost improvements every six months (current baseline versus the new cost target) and continue to become more and more skilful in reducing by using the kaizen improvement methodologies and approaches.

One example is 'activity value analysis kaizen' to review the processes handled by the staff. Analyse every activity performance by every staff in your organisation. Learn how to create efficiencies in every activity by eliminating non-value-adding activities, combining activities, and optimising those activities. Reducing the number of activities or processes only then will result in resource optimization and continue to deliver higher productivity. This must be part of the CC team's methodological approaches to optimise cost.

Safety First

A Good Safety Behaviour Is Everything
Special article by Dr Azlan Nithia (2024)

Importance of Safety Behavior in Manufacturing Organisations

Safety is a **top priority** for manufacturing companies for several reasons when comes to competing globally.

1. **Legal compliance.** Manufacturing companies must adhere to various safety regulations and standards set by local, national, and international governing bodies. Failure to meet these requirements can result in legal consequences, such as fines, penalties, or even business shutdowns.

2. **Reputation and brand image:** Companies that prioritise safety demonstrate their commitment to the well-being of their employees, customers, and communities. This dedication helps build trust and a positive brand image, which can be a competitive advantage in global markets. Customers are more likely to choose products from a company known for its safety measures.

3. **Employee retention and productivity.** A safe work environment instils confidence in employees, boosts morale, and improves job satisfaction. When employees feel safe and valued, they are more likely to stay with the company and be productive. High employee turnover due to safety hazards can disrupt operations, increase training costs, and negatively affect overall productivity.

4. **Reduce cost.** Implementing effective safety measures can help prevent accidents, injuries, and equipment damage. This reduces costs associated with medical expenses, workers' compensation claims, insurance premiums, and machinery repair or replacement. Safety-conscious companies can utilise their resources more efficiently, resulting in increased profitability and lower production costs.

5. **Global compliance**. Many companies aim to expand their operations globally. However, different countries have their own safety regulations, and complying with these requirements is critical for market entry. Ensuring safety across all locations allows manufacturers to meet international standards and gain access to new markets without facing legal or regulatory barriers.

In conclusion, prioritising safety allows manufacturing companies to maintain legal compliance, enhance their reputation, retain employees, reduce costs, and enter global markets more effectively. By ensuring the well-being of their workforce and stakeholders, companies set themselves up for long-term success in the competitive global manufacturing industry.

Safety Is a Behavior and Starts with a Good 5S Program

Safety in manufacturing companies is more than just a set of regulations or norms; it involves action that has been thoroughly engrossed as part of the corporate spirit and individual psychology. Doing so necessitates a proactive, practical attitude and constant effort. An organisation that provides a safe work environment by clearing slips, trip hazards, and cleanliness takes the highest priority (Ladewski and Al-Bayati, 2019). In addition, through the means of 'standardise' and 'sustain', these 5S practices become part of an organisational culture to be passed along from generation to generation so as not only to promote consistent safe conduct but also to contribute towards endowing meaning in life itself.

Safety starts with a systematically deep commitment for the 5S culture in the organisation.

Through the 5S program, a culture that values safety among all employees can be developed. It provides a systematic framework that not only organises the work environment but also creates an atmosphere of order and vigilance with respect to safety procedures (Muotka et al., 2023). When people seriously undertake the application of 5S, they strengthen safety as behaviour and make it an integral part of their daily work.

How to Create a Good Safety Behavior in the Manufacturing Industry?

A good safety culture in the manufacturing industry requires a comprehensive strategy including education, empowerment, awareness, and persistent reinforcement. A safety culture starts at the top. Safety is put first in everything the organisation does or carries out no less than leadership commitment. This can be measured by actively promoting safety and allocating truly committed resources to improve it, as well as integrating a safety focus into the mission and values of an organisation (Butt, 2020). Safety behaviour is a key aspect of training, education, and practice. An intensive training program should go beyond the necessary safety precautions and procedures so as to base them on a sound rationale (Muchiri et al., 2019). The more employees understand the reasons underlying such measures, the easier it will be for them to internalise and apply safety principles in their daily work, provide a safe workplace, operate safe equipment, and importantly look out for each other's safety in the organisation by stopping any unsafe acts.

They have to create transparent channels of communication that allow all employees (including cleaners and operators) to report possible hazards or make suggestions without fear of reprisal, strengthening the safety culture. Create a recognition and reward system to appreciate individuals and teams frequently. Let the employees come forward to tell the leadership team about their story of safety and activities being improved in their area.

Involvement and empowerment promote safety behaviours (Butt, 2020). There is a sense of responsibility and proprietorship among personnel who are involved in safety committees or task forces. When employees are allowed to actively participate in identifying dangers, drafting responses, and implementing enhancements for safety problems, a sense of collective responsibility towards the problem of ensuring safe working conditions develops among the staff as well. Consistency is the secret to successfully reinforcing safe behaviours (Ladewski and Al-Bayati, 2019). Feedback mechanisms, inspections, and audits act as good checkpoints to evaluate the efficacy of safety procedures and pinpoint possible points needing improvement. A monthly recognition and reward system for the most improved and good safety performing areas creates deep individual and team commitments to improving safety. Remember never ever use safety reasons to punish any employees because it will always drill upwards towards the leader.

How Great Companies Deploy the Safety Champions Program Successfully?

The safety champions program is an effective method used by prominent organisations to promote a culture of safety throughout their establishment. The safety champions are the representatives. Through deeds and promotion, they demonstrate to their colleagues a dedication to safety (Tahaei et al., 2021). Safety champions should be chosen on the basis of their enthusiasm, influence, and willingness to promote safety; this will help ensure the program's success. After selection, these champions get specialised training that instils in their deep knowledge of safety procedures efficient ways to communicate and leadership abilities. Through this instruction, they become change agents inside their respective groups and work areas. They help to set standards and get everyone directly involved in discussions concerning safe working conditions (Tahaei et al., 2021). These advocates are often supported financially or have opportunities for career growth all along their career by these prominent corporations, giving them the necessary tools to effectively advocate and promote safety in whatever field they work. Communication is an essential ingredient in any safety champions' campaign.

Leading organisations ensure that these advocates receive regular forums to share success stories, best guidance, and difficulties encountered regarding the execution of safety programs. These advocates encourage constructive discussions, drawing in the input and recommendations of team members (Muchiri et al., 2019). This is a program that effective companies implement by cultivating a safety champions' network with full commitment, a fully involved leader (the number one leader) with this program, and adequate equipment and support through efficient communication channels. These advocates are an integral part of building a safety-oriented environment throughout the entire organisation (Tahaei et al., 2021). They affect employee behaviour, promote cooperation, and influence continuous improvements in standards for workplace safety.

Why Safety Champion Programs Fail and Difficult to Sustain?

Although potentially beneficial to an organisation, in the long run, safety champion programs might encounter challenges as well. A notable factor is unclear goals and top-level executive support, which might end up in failure (Sánchez-Gordón and Colomo-Palacios, 2020). If the leadership does not make the objective of the program clear or when there is insufficient time and training resources to throw at it, safety champions will frequently find themselves unable to bring about big changes (Ladewski and Al-Bayati, 2019). The phenomenon of lacking official support and investment might make the programs unable to function effectively. Additionally the difference between advocates and a general labour force can stand in the way of these programs (Sánchez-Gordón and Colomo-Palacios, 2020). Discontented peers may find the influence of safety champions constrained. Ultimately factors such as cultural barriers or the mistrust between bosses and employees make it difficult for safety champions to explain why they should care about workers' safety.

Why Investigating a Safety Incident in a Manufacturing Company Is Very Important

Investigating safety incidents in manufacturing companies is crucial for several reasons:

1. **Identifying the root causes.** Conducting an investigation allows the company to determine the underlying causes of the incident. By identifying the root causes, the company can implement effective corrective actions to prevent similar incidents in the future.

2. **Preventing recurrence.** Understanding why an incident occurred helps in developing and implementing preventive measures. By addressing the root causes, manufacturing companies can take steps to eliminate or minimise hazards, improve safety protocols, and provide additional training to employees to prevent similar incidents from happening again.

3. **Ensuring compliance.** Investigating safety incidents ensures that the manufacturing company remains compliant with regulations and standards set by regulatory authorities. Compliance is essential not only to avoid penalties but also to maintain a safe working environment for employees.

4. **Protecting our employees.** Our employee safety is of paramount importance. Investigating safety incidents demonstrates the company's commitment to the well-being and protection of its employees. By thoroughly investigating incidents, a company can take proactive measures to create a safer work environment and prevent injuries or harm to its workforce.

5. **Building trust and morale:** When manufacturing companies investigate safety incidents promptly and transparently, it helps build trust among employees. By acknowledging and addressing incidents, companies show that they prioritise the safety and well-being of their workforce. This, in turn, can boost employee morale and engagement.

6. **Reducing costs.** Safety incidents can have significant financial consequences for manufacturing companies, including increased medical costs, worker's compensation claims, equipment damage, production delays, and potential legal actions. Investigating incidents helps identify areas where costs can be reduced through improved safety procedures, training, and risk mitigation.

Overall investigating safety incidents in manufacturing companies is critical to protect employees, prevent future incidents, maintain compliance with regulations, and create a positive and safe work environment.

What Are the 10 Steps for a Good Safety Incident Investigation Process in a Manufacturing Company?

When conducting a safety incident investigation in a manufacturing company, it is important to follow a systematic process to ensure a thorough and accurate investigation. Here are ten steps to consider:

1. Secure the scene. As soon as an incident occurs, make sure to secure the scene to prevent further accidents or tampering with evidence. Restrict access and preserve any physical evidence.

2. Ensure safety. Prioritise the safety of personnel during the investigation. Make sure that everyone involved is safe and provide necessary medical attention if required.

3. Form an investigation team. Assemble a team with the necessary expertise to conduct the investigation. This team may include safety professionals, supervisors, engineers, and other relevant personnel.

4. Gather information. Collect all available information related to the incident, including witness statements, photographs, videos, and any relevant documents or records.

5. Conduct interviews. Interview all individuals involved in or witnessing the incident. Ask open-ended questions to gather detailed information about the sequence of events and any contributing factors.

6. Analyse data. Analyse the collected data to identify the immediate, underlying, and root causes of the incident. Look for trends, patterns, and any deviations from normal procedures.

7. Determine corrective actions. Based on the identified causes, develop appropriate corrective actions to prevent similar incidents from happening in the future. These actions should

focus on addressing the root causes rather than just surface-level issues.

8. Implement corrective actions. Once the corrective actions are determined, create a plan for implementation. Assign responsibilities, set timelines, and develop a monitoring process to ensure effective implementation.

9. Review and update policies. Review existing safety policies, procedures, and training programs to identify any gaps or deficiencies that may have contributed to the incident. Update these policies as needed.

10. Communicate findings. Share the investigation findings, including the identified causes and recommended corrective actions with relevant stakeholders. This communication helps create awareness and transparency and facilitates organisational learning.

Remember every incident is unique, and the investigation process may need to be tailored accordingly. These steps provide a general framework to ensure a comprehensive and effective investigation.

What Are the Difficulties Faced by Manufacturing Companies while Investigating a Serious Safety Incident?

Manufacturing companies may face several challenges when investigating serious safety incidents:

1. Limited resources. Manufacturing companies may have limited resources, both in terms of personnel and financial resources, which can make it challenging to allocate sufficient time and effort to the investigation process.

2. Complexity of operations. Manufacturing operations can be complex, involving multiple machinery, equipment, processes, and interactions between workers. Investigating a serious safety incident may require a deep understanding of these complexities to identify the contributing factors accurately.

3. Time constraints. Production schedules and delivery deadlines are critical for manufacturing companies. Investigating a safety incident may disrupt operations and lead to additional costs and delays, making it difficult to allocate the necessary time for a thorough investigation.

4. Pressure to resume operations. In cases where an incident leads to a halt in production, there may be significant pressure from stakeholders to resume operations as quickly as possible. This pressure can lead to a rushed investigation or shortcuts, which may compromise the quality and accuracy of the investigation.

5. Lack of expertise, Conducting a comprehensive safety incident investigation requires specialised knowledge and skills. Manufacturing companies may face challenges in finding internal investigators who possess the necessary expertise or in hiring external experts to assist with the investigation.

6. Fear of legal and reputational consequences. Serious safety incidents can have significant legal and reputational implications for manufacturing companies. There may be a fear of negative publicity, regulatory penalties, or potential litigation, which can influence the transparency and objectivity of the investigation.

7. Employee cooperation. Obtaining accurate and honest information from employees involved in the incident or witnesses can sometimes be challenging. Fear of disciplinary action, job loss, or reprisals can impact their willingness to cooperate fully, making it difficult to gather all the relevant information.

References

Butt, J. (2020) 'A strategic roadmap for the manufacturing industry to implement industry 4.0', *Designs*, 4(2), p. 11.

Ladewski, B. J. and Al-Bayati, A. J. (2019) 'Quality and safety management practices: The theory of quality management approach', *Journal of Safety Research*, 69, pp. 193–200.

Muchiri, M. K., Mc Murray, A. J., Nkhoma, M., and Pham, H. C. (2019). 'How transformational and empowering leader behaviors enhance workplace safety: A review and research agenda', *The Journal of Developing Areas*, 53(1).

Muotka, S., Togiani, A., and Varis, J. (2023) 'A Design Thinking Approach: Applying 5S Methodology Effectively in an Industrial Work Environment', *Procedia CIRP*, 119, pp. 363–370.

Sánchez-Gordón, M., and Colomo-Palacios, R. (2020, June) 'Security as culture: a systematic literature review of DevSecOps'. In *Proceedings of the IEEE/ACM 42nd International Conference on Software Engineering Workshops*, pp. 266–269.

Tahaei, M., Frik, A., and Vaniea, K. (2021, May) 'Privacy champions in software teams: Understanding their motivations, strategies, and challenges'. In *Proceedings of the 2021 CHI Conference on Human Factors in Computing Systems*, pp. 1–15.

CHAPTER 3

Growth and Competing Globally

Becoming Globally Competitive

The term globally competitive refers to the ability of organisations to produce goods and services that can compete effectively with those produced by other countries on a global scale. To be able to compete effectively in the global marketplace in every aspect of manufacturing (productivity, efficiency, innovation, quality, etc), organisations have to deliver good quality products, globally competitive costs, and the ability to respond rapidly (agility) to any market changes.

Becoming globally competitive in manufacturing requires a comprehensive and holistic approach that encompasses various aspects of the manufacturing ecosystem. By focusing on important key factors, manufacturing companies can position themselves to thrive in the global market and drive growth and development.

Let's understand the possible meanings of becoming globally competitive:

1. **Enhancing economic growth**. A key meaning of becoming globally competitive is the ability to foster economic growth by effectively participating in global markets, attracting foreign investment, and increasing exports.

2. **Expanding market reach.** Being globally competitive means reaching a broader customer base by expanding into international markets and diversifying revenue sources.

3. **Increasing productivity.** Becoming globally competitive entails improving productivity levels through innovation, technology adoption, and efficient business processes to match or surpass global standards.

4. **Fostering innovation.** A crucial aspect of global competitiveness is the ability to innovate and develop new products faster with efficient services, and reliable processes that meet global market demands and stay ahead of your competitors.

5. **Developing a skilled workforce.** Becoming globally competitive requires nurturing a highly skilled and adaptable workforce that can quickly adapt to meet the demands of the global marketplace and drive competitive economic growth.

6. **Building strong international partnerships.** Global competitiveness involves establishing partnerships and collaborations with international entities, fostering knowledge exchange, optimising the borderless global economy, and leveraging cross-border synergies.

7. **Embracing globalisation.** Being globally competitive means recognising and adapting to the interconnectedness and interdependence of economies, cultures, and societies in the globalised world.

8. **Investing in research and development.** Building a strong research and development infrastructure is crucial for achieving global competitiveness as it enables continuous innovation, technological advancements, and constantly realigning market relevance.

9. **Ensuring regulatory and policy framework.** A supportive regulatory and policy environment that fosters competition protects your intellectual property, encourages entrepreneurship within your organisation, and facilitates trade optimisation is necessary for global competitiveness.

10. **Prioritising sustainability and environmental responsibility.** Becoming globally competitive also means adopting sustainable

practices and trends, minimising environmental impact, and meeting global standards for responsible and evidence of ethical business conduct.

Strategies for Creating Growth and Competitiveness

1. **Research and identify market opportunities**. Conduct market research to understand current trends in your industry, demands, and potential gaps in the manufacturing industry (which you can go get). Identify emerging sectors and areas where your company can excel and expand. Review the current product profile and diversify your current product profile into a wider range of product lines and options.

2. **Improve product development and innovation**. Invest in research and development resources to enhance existing products or create new ones. Encourage innovation within your organisation by fostering a culture that values creative thinking and rewards employees for suggesting and implementing new ideas.

3. **Enhance operational efficiency**. Optimise your manufacturing processes to improve productivity and reduce costs. Streamline workflows, invest in automation technologies, and implement lean manufacturing principles and value stream mapping to eliminate waste and increase operational efficiencies in all functions.

4. **Expand your customer base**. Develop a comprehensive marketing and sales strategy to target new customers and expand into new product profiles. Consider expanding into new geographical regions or diversifying your customer base by targeting different market segments while not complicating the manufacturing processes. Consider adding any new process capabilities as needed basis in your manufacturing operations so that you can expand the product profile offerings.

5. **Build strong partnerships**. Collaborate with suppliers, distributors, and other stakeholders to strengthen your supply chain and improve access to resources. Joint ventures or strategic

alliances with complementary businesses can also lead to new growth opportunities.

6. **Invest in employee training, skills, and talent development**. Continuously upskill and train your workforce to keep up with the latest industry advancements. Provide opportunities for employees to enhance their skills and knowledge, fostering a culture of continuous learning and improvement and expanding their talent capabilities.

7. **Embrace digital transformation and connected factory**. Adopt advanced manufacturing technologies such as robotics automation, artificial intelligence, and the internet of things (IoT) to improve processes, increase productivity (intensify the program 90 + 7), and enable data-driven decision-making. To grow the manufacturing capabilities, it is also important to implement the connected-factory strategies so that all the machines, processes, assembly, and shipping are digitally connected and driven by real-time dashboard updates that trigger rapid and accurate corrective actions.

8. **Focus on sustainability**. Incorporate sustainable practices into your manufacturing processes to reduce environmental impact, and meet the growing demand for eco-friendly products and processes. Increase recycling, recycle processed water, embark on zero-waste programs, and reduce energy consumption. This can also attract environmentally conscious customers and improve your organisational brand reputation and commitment.

9. **Explore government support and incentives**. Seek any available government programs, grants, incentives, and tax breaks that support manufacturing growth. These can include research and development grants, technology support, export assistance programs, and funding for workforce skill development.

10. **Monitor and adapt**. Regularly review and analyse key performance indicators (KPIs) to track your daily and weekly progress. Stay informed about industry trends, competitions, customer preferences, and changes in regulations to adapt and pivot your strategies as needed.

Remember that implementing these strategies may require leadership team commitments, careful planning, resources (people and processes), and time. Keep a long-term perspective and be open to evolving your approach as you navigate the challenges and opportunities that arise in the related manufacturing industry and your competition.

Strategies to Make Manufacturing Plants Globally Competitive

There are several key strategies and practices they can implement:

1. **Embrace technology.** Invest in advanced manufacturing technologies such as automation, robotics, artificial intelligence, and the internet of things (IoT) to improve productivity, reduce costs, and enhance quality. Implementing advanced data analytics systems can also enable real-time monitoring and predictive maintenance.

2. **Implement lean manufacturing principles.** Adopt lean manufacturing principles and practices such as just-in-time (JIT) manufacturing, develop continuous improvement as an organisational culture, and waste reduction to increase efficiency as a relentless journey towards perfection, reduce lead times (never stop cutting lead times in every process), and optimise resource utilisation (doing more and more with less and less).

3. **Focus on quality and safety.** Establish a robust quality control system to ensure that products meet international standards and customer expectations. Implement quality management practices such as Six Sigma and total quality management (TQM) to improve product reliability and customer satisfaction.

Establish a plantwide safety program, implement the safety champions program throughout your organisation and give visible strong leadership support. Make safety the first, last and always in everything your workforce does in the organisation.

4. **Develop a talented skilled workforce.** Invest in employee training and development programs to enhance the skills and knowledge of your workforce. Encourage cross-training and upskilling initiatives to have a versatile workforce capable of adapting to changing technologies and processes rapidly and proactively.

5. **Foster innovation and R and D.** Allocate resources for research and development activities to drive innovation in product design, manufacturing processes, and materials. Encourage collaboration with universities, technical internship programs, research institutions, and other industries to stay at the forefront of technological advancements.

6. **Establish an efficient global supply chain network.** Build strong relationships with suppliers and ensure a resilient supply chain to minimise disruptions. Develop a supplier evaluation and qualification process to ensure high-quality inputs at competitive prices. Constantly reduce lead times from your order confirmation, manufacturing and delivery, the end-to-end lead times.

7. **Implement sustainable practices.** Adopt sustainable manufacturing practices to reduce environmental impact and attract environmentally conscious customers. This includes energy efficiency, waste reduction, recycling, and using eco-friendly materials.

8. **Stay agile and flexible.** Develop a culture of adaptability and agility to quickly respond to changing market demands and customer preferences. Employ flexible production systems that can easily scale up or down based on market demand fluctuations with quick change-over systems in every manufacturing process.

9. **Develop strategic partnerships and alliances.** Collaborate with industry partners, research institutions, and technology providers to leverage their expertise, share resources, and achieve mutual benefits. This can help to find ways to reduce costs, accelerate innovation, and open new market opportunities or new product profiles.

10. **Understand global market trends.** Stay updated on global market trends, emerging technologies, and regulatory changes to anticipate future demand and align production strategies accordingly. Develop a thorough understanding of target markets, including cultural preferences, local regulations, and competitive landscapes.

By implementing these strategies, manufacturing plants can enhance their competitiveness on a global scale, increase market share, and sustain growth in the long term.

Leadership Failures That May Cause Manufacturing Plant Closures

There can be various reasons why manufacturing company leaders fail to stop the plant closure. Here are some possible factors that could lead to those failures, which include:

1. **Lack of strategic planning.** If leaders do not have a well-defined long-term strategy for the company, they may struggle to find ways to prevent closures. This could include not adapting to changing market trends or failing to anticipate potential challenges.

2. **Inefficient cost management.** Manufacturing closures often occur due to financial difficulties. If leaders fail to effectively manage costs and control expenses, it can lead to financial instability and ultimately closures.

3. **Failure to innovate.** In today's rapidly changing business environment, innovation is crucial for manufacturing companies to remain competitive. If leaders fail to invest in research and development, adapt to new technologies, or identify new market opportunities, it can result in closures.

4. **Lack of foresight.** Leaders who do not have a clear understanding of market dynamics, lack understanding of the competitors, or fail to anticipate industry changes may find it challenging to

prevent closures. This includes not staying updated on industry trends, emerging technologies, or evolving customer demands.

5. **Inadequate talent management**. One of the key resources for any manufacturing company is its workforce. If leaders do not attract, retain, and develop skilled employees, it can lead to a decline in productivity, poor product quality, or an inability to meet customer demands, ultimately leading to closures.

6. **External factors**. Sometimes, closures may be out of the control of manufacturing company leaders, but this is not always a common reason for closures. Economic downturns, changes in government regulations, trade barriers, or shifts in consumer preferences can all impact the viability of manufacturing businesses.

7. **A wrong leader**. It is important to note that the elected leader of the organisation must be passionate and wholeheartedly committed to the manufacturing site and enjoy being with the people to drive transformation and sustainable competitiveness.

It is important to note that these are general factors, and each situation may have its own unique combination of challenges.

Take Quick Action on Poor-Performing Employees

Taking quick action on poor-performing employees in a manufacturing company is important to maintain productivity and ensure the overall success of the business. Delaying action on poor performance will only worsen the situation for the company and negatively impact the other performing leaders. You must take quick action.

Here are some steps you can take.

1. Identify the issues. Clearly define the poor performance issues displayed by the employees. This could include low productivity, missed deadlines, frequent errors, or a negative attitude.

2. Provide feedback and set expectations. Schedule a meeting with the employee to discuss their performance, highlighting

the areas that require improvement. Clearly communicate the expectations and standards that need to be met.

3. Offer support and resources. Identify any training or resources that could help the employee improve their skills or knowledge. Provide them with guidance and support to help them meet the required performance levels.

4. Monitor progress. Regularly monitor the employee's performance after providing feedback and support. Keep track of their progress and provide continuous feedback.

5. Take disciplinary action if necessary. If the poor performance continues despite support and feedback, you may need to take disciplinary action. Follow your company's established procedures and guidelines for this. It could range from verbal or written warnings to more severe consequences such as suspension or termination.

The timing of taking action depends on the severity and impact of the employee's poor performance. If their performance is significantly affecting the productivity, safety, quality or overall functioning of the manufacturing process, it may be necessary to take quicker action. However, it is important to give employees a fair opportunity to improve and provide the necessary support before resorting to any disciplinary action.

Improving the Poor Performing Leaders

Improving poor-performing leaders in manufacturing companies requires a combination of strategies to address their shortcomings and develop their skills. Here are ten effective ways to achieve this:

1. Clearly define expectations. Ensure that leaders have a clear understanding of their roles and responsibilities, as well as the performance expectations that accompany them.

2. Provide regular feedback. Establish a feedback loop where leaders receive constructive criticism and guidance on their

performance. Regularly assess their skills and areas in need of improvement.

3. Offer leadership training. Invest in leadership development programs tailored to the specific needs of the manufacturing industry. These programs should focus on areas such as communication, problem-solving, decision-making, and team building.

4. Encourage mentoring and coaching. Assign mentors or coaches to poor-performing leaders, empowering them with additional support and guidance. Mentors can share their expertise and experiences, helping the leaders develop new skills.

5. Foster a learning culture. Promote a culture of continuous learning within the organisation. Encourage leaders to seek out new knowledge, attend workshops or conferences, and stay updated on industry trends.

6. Team-building activities. Engage leaders in team-building activities that promote collaboration and strengthen relationships with their subordinates. Encouraging open communication among team members can lead to improved leadership performance.

7. Provide resources and tools. Ensure that leaders have access to the necessary resources, tools, and technology to perform their roles effectively. This includes providing adequate training on new systems and processes.

8. Set achievable goals. Collaborate with poor-performing leaders to set SMART (specific, measurable, achievable, relevant, time-bound) goals that align with the overall objectives of the organisation. Regularly monitor progress and provide support as needed.

9. Develop emotional intelligence. Emphasise the importance of emotional intelligence in leadership. Provide training and support to help leaders become self-aware, manage their emotions, and build positive relationships with their teams.

10. Hold leaders accountable. Establish a system where leaders are held accountable for their performance. Ensure there are

consequences for poor performance while also recognising and rewarding improvements.

Remember every leader is unique, so a tailored approach is crucial to address their specific needs, shortcomings, and challenges.

Make the Right Things Happen

In manufacturing nothing happens by itself, everything is made to happen by you or by someone in the organisation whether it is the right or the wrong thing. They are all made to happen. We as a leader had made it happen for the good or the bad. We had allowed it to happen. Always the leader must set a clear direction, create and make good (or the right) things happen. The good or the right things do not happen by themselves in a manufacturing organisation. They are created by the actions of our people.

You must make the right things happen whether in life, in manufacturing, or in anything. If you want to lose some weight, you have to make it happen. Losing weight does not happen by itself. When you are upset about a certain situation, you could lose your anger with your people, but that is wrong. How to ensure we don't lose our anger, we must make it happen so that we do not lose our anger.

Remember only wrong things may happen automatically (even though it could be the result of a certain action), but right things, you must make them happen. They don't happen by themselves (the right things). Even to say sorry is the right thing to do, but we must make it happen by saying the word sorry.

It is all about making it happen. Get it done by using the 'three rights'.

The right thing, the right way, and at the right time.

When you are making things happen, obviously we always expecting good results or positive outcomes. Ask the below the 'three rights':

- *Am I doing the right thing* to make the right thing happen?
- *Am I doing it the right way* to make it the right way happen?
- *Am I doing it at the right time* to make it happen at the right time?

It is always important that we train our people to ask the above three rights questions before moving forward with any of our actions to make the right things happen in the right way and at the right time.

Building Trust with Your People Is Important

It is critical for the leader and everyone in the leadership team to know and acknowledge the importance of *trust*. It is super important especially when an organisation is in a manufacturing crisis and in the process of transforming to become a competitive manufacturing operation. Trust will unite or break the team's unity and belief.

Trust is everything! Trust is difficult to build and very easy to lose!

We, as leaders, it is crucial to be the first to show our people confidence and motivate them to act in the same way. Listening to our people is a form of showing respect for them and their ideas. The more we listen and respect our people, the more we build trust with them. To trust us, it is important that our people feel they are in a safe zone. Therefore, leaders must create that safe zone or space for all our employees in the organisation. This means that we as leaders must never have hidden agendas with our people. Show integrity by having open conversations to achieve a common shared principle.

When you do not trust others, then you make others not trust you too. Seriously people who cannot trust others will fail to be leaders because they will not be able to delegate almost anything, or they will end up doing the overcontrolling of everything they delegate.

Everyone in the organisation must see and feel the persistence of what we as leaders say, and every leader must be 'walking the talk'. We as leaders cannot say things, and then we do not demonstrate what we have said or been saying to our people. It is important for all to know, never ever underestimate our people's intelligence. Our people can and will be constantly reading our minds and our intentions, so always be aware and be conscious of this.

If ever the trust is broken with our people, it is very difficult to build back the broken trust with our people. Sometimes it is almost impossible. It is also very true in our personal life, especially with our friends and family, the same as it is at work with our employees.

Always remember to keep the promises we make to our teams and employees. Tell the truth, say what you want to say with full honesty, and make everyone feel that it is truly coming from your heart (never underestimate the intelligence of your people). Never ever create any doubts in the minds of our employees regarding our honesty.

Sometimes due to certain difficult times or situations, we as leaders could be easily swayed to do or even to say things that are different from real situations versus what we had promised before to our teams and employees. This is extremely dangerous. Do not do it. Only say the right thing, the right way and at the right time. If the time is not right, then don't say anything. You must wait for the right moment and do it when the time is right to say those right things in the right way.

Always be persistent in our behaviour during very challenging times or situations. It can be very difficult but it is important.

We must demonstrate that we are interested in showing consideration and sensitivity to the needs and interests of our people by giving timely support. At all times, demonstrate and act in a way that will protect the interest of our people and show that we are interested in their safety and security at work as an example. Show that you are not for self-gains and exploiting anyone at work, but you are working with everyone towards a common goal.

Every time you communicate, be precise with what you are saying, say the exact things you want to communicate. Then explain your rationale for the decisions you are suggesting. Make all feel your openness and allow everyone to participate in the decision you and your leadership team are making, providing a way for the employees to voice out their concerns.

Through trust, we as leaders can create a sense of belonging. As a collective team, we become the 'one team' with a common identity and with shared values.

It is equally important to know and understand what erodes our trust!

**It is always about them (our employees)
and not about what it is for me.**

I have heard many times—divide and rule the teams to achieve our goals. Please, this is very wrong. It does not apply at work. Never ever do this. It is very harmful because it quickly erodes the trust we may have in our employees and teams. As I had mentioned before, never underestimate our people's intelligence.

The six things of 'do nots' of building trust:

1. Do not blame your employees or staff for their mistakes and demonstrate unfair accusations. Always focus on the problem or the issue and never on the person. Respect the person with the problem and solve the problem collectively.
2. Do not show favouritism and discrimination in any way or in any form. Show respect to all, show that you are a leader for all, and treat everyone's feedback seriously.
3. Do not break the promises that you made to your people. You must always walk the talk and you must be seen doing the things that you preach to others.
4. Do not disclose or misuse anyone's personal information without getting prior approval from the person involved. Respect everyone's confidentiality.

5. Do not get involved in any form of gossip in any form about any of your employees. Discuss any gossip you may hear with the person face to face and not behind the person's back.
6. Do not encourage competition that will erode teamwork or people collaborations. Focus on common goals that drive 'one team, one goal'.

It is important that we as leaders create a sense of shared values, shared destiny, and shared purpose and build trust by walking the talk of all these. Leadership is never about imposing onto our employees, but it is about developing or creating a sense of purpose and common destiny—knowing, committing, and living it! As leaders, we must know how to express ourselves completely, we must know ourselves, and we must know our abilities and our shortcomings. Leadership may be difficult to define, but it is easy to recognise if you see it. Therefore, it is always being honest through which we constantly influence our people positively to achieve our common goals.

Develop T Leaders in the Organisation

A 'T leader' typically refers to someone who possesses deep expertise or knowledge in a specific area (the vertical line of the 'T') while also having the ability to collaborate and work effectively across various functions or disciplines (the horizontal line of the 'T'). This combination allows T leaders to excel in their specialised area while also understanding and appreciating the broader organisational context.

T leadership emphasises the importance of both specialisation and cross-functional collaboration. These leaders bring their specialised skills, experience, and knowledge to solve complex problems and make informed decisions in their field. Simultaneously, they have the ability to communicate, collaborate, and integrate their expertise with other functions or departments to achieve overall organisational goals.

T leaders are often valued for their ability to bridge gaps between different teams or departments, facilitating communication, cooperation, and a shared understanding. This type of leadership can be particularly effective in organisations that require a balance between deep expertise and collaboration across functions, fostering innovation, and driving overall performance.

We need T leaders in our organisation, the leaders who can take on leadership responsibilities for multiple functions and are able to deliver sustainable good performance results. This requires leaders who are eagerly self-driven towards achieving better results. This type of leader wants to do more and is energised to deliver better performance all the time on every assignment given to them.

It is important that we are constantly able to find and recognise these types of leaders in our organisation. Reward them, expand their job responsibilities, and give them challenging assignments. Multiply this type of leadership throughout your organisation to achieve a high-performing organisation culture.

CHAPTER 4

Innovation and Leadership Strategy—Engaging Our People

Innovations are commonly seen as an individual outcome. There would be certain individuals in an organisation who look up to an innovative or creative person. In this case, the innovation remains as an individual outcome. The organisation has an option, depending on its interest to develop innovation and creativity as an organisational culture or a team culture. To develop an innovative culture, the organisation requires the right leader to develop and prosper this culture. It cannot be just an individual playing a magical role.

Therefore, to cultivate an innovative culture, we require good leadership. Only the right leader can create a culture of innovation. Many times, innovations fail due to poor leadership. The biggest challenge is to find the right leader to lead this innovative team. Similarly poor leadership kills innovation in the organisation or amongst the teams. We need to ensure the organisation identifies the right leader, creates challenges, promotes innovation, and rewards innovation.

Old Ways Will Not Create New Ways

The old generations will continue to create old generational ways. New generations cannot be created by the old generation (this is my personal opinion based on my experience). Similarly old methods will result in old outcomes. To get new results, we need new methods or new ways, and new ways will deliver new outcomes or new results. We need new thinking for new outcomes. The old generation will always remain the old generation if they are not getting updated with new generation ideas. For the old generation to create the new generation's ideas, it is possible only by having constant transformational learning to distance itself from the old generation's thinking or ways. These are those who are constantly learning from the new generation or the younger generation and are comfortable with the change and practice new generation ideas just like the younger generations.

Creating a Successful Innovation Leadership Strategy in a Manufacturing Organisation

Special Article by Azlan Nithia (2024)

Introduction

Technology has revolutionised the way organisations carry out their activities in the manufacturing sector. Technology has compelled businesses to embrace innovation leadership to smoothly shift from traditional operational methods, where consumers had to physically visit stores to make purchases, to modern business practices driven by innovation. This transition aims to enhance the organisation's profitability, establish its brand reputation in the global market, and expand its customer base. By employing forward-thinking leadership, a company can implement a business strategy that fosters a culture of acknowledging team members' contributions and integrating their ideas into company projects, ultimately driving the organisation's success. To establish an innovation leadership strategy within the firm, leaders must comprehensively understand the always-evolving consumer demands, foster trust among employees, and encourage diversity. Hence, implementing a leadership innovation strategy is significant in the manufacturing industry.

Importance of Creating a Successful Innovation Leadership Strategy

1. Adapting to Changing Consumer Needs

Gaining insight into consumer requirements and preferences is crucial to preventing competitive enterprises from outperforming by attracting consumers to buy their products. The organisation's continuous innovation enables it to thoroughly analyse the requirements of its consumers and develop products that effectively cater to their needs by aligning them with their target audience (Aslam et al., 2023). Income, age, emerging trends, peer pressure, and societal changes are just a few of the variables that affect consumer demands and preferences (Aslam et al., 2023) Through acquiring knowledge and comprehending consumer demands, the firm can augment its sales by fostering loyalty among its current customers and improving customer happiness, thus attaining a competitive edge. Satisfied customers who are pleased with the company's products and customer service tend to bring in new customers by speaking well about the business when interacting with their peers or posting on social media (Aslam et al., 2023). Testimonials from previous customers are crucial. They enable the business to maintain its relevance in the market and develop itself as a prominent leader in the manufacturing industry.

2. Cultivate trust among employees.

Innovative leaders can establish and sustain trust with their subordinates, so cultivating a favourable working environment that nurtures employee creativity and efficiency ultimately contributes to a business's success. Trust is a crucial and effective tool for fostering unity among employees. It enables them to rely on their leaders for support and collaborate as a team to complete projects within specified time frames successfully. This commitment goes beyond individual interests and reflects a shared dedication to the organisation's values and objectives (Mitcheltree, 2021).

Innovative leaders can apply several tactics to establish employee trust, such as fostering transparent communication and enabling team members to express their thoughts freely without apprehension of reprisal. In addition, fostering an environment where employees are encouraged to express their opinions and provide input regarding a particular project or topic instils a sense of worth. It fosters a sense of inclusion, establishing faith in their leaders' ability to steer them toward accomplishing the established objectives (Mitcheltree, 2021). By consistently and reliably displaying their actions, innovative leaders can establish trust among employees. Employees can place their trust in leaders who fulfil the commitments they made to them when assuming the leadership role. Leaders can assess their competence and inclination to steer others towards achieving company objectives effectively. Trust is fostered among the workforce when leaders acknowledge the various capabilities of each person and assign tasks based on their skills and experiences (Mitcheltree, 2021). Employees believe their leaders should alleviate job stress by delegating tasks to them that they can handle and by recognizing their contributions.

3. Enhance operational efficiency.

In a manufacturing organisation, operation efficiency plays an essential central role in determining the success and profits of the organisation as it influences different aspects of the business. Enhancing production, which is a function of operational efficiency, is essential for lowering operating costs by making the best use of available resources and reducing waste, which increases product output. Innovation leaders are critical in this process as they drive operational excellence. Through their guidance, they ensure employees utilise the resources available during production and streamline workflow by allocating work responsibility, which enhances production. Innovation leaders can foster a culture where operational efficiency becomes a shared responsibility by instilling a commitment to improvement at every level of the organisation.

4. We are empowering employees.

Those in leadership are responsible for utilising innovation to promote creativity and professional growth among employees of an organisation.

Leaders can encourage employees to think beyond the information in their projects, ensuring they add value to the project and foster a sense of pride in their careers. Innovative leaders can promote autonomy for their employees by leaving them to make decisions concerning some of the individual projects they are handling. Employees appreciate when their leaders trust them enough to make the correct decision and contribute to the organisation's success without monitoring them for every step they take in completing their projects.

5. Encourage the adoption of a growth mindset among employees.

Additionally, in promoting innovation among employees, workers need to be encouraged to adopt a growth mindset. Having such a mindset enables the employee to overcome the challenges they face within their work environment since they are confident in taking risks or learning new things to incorporate into their work. Leaders who promote a growth mindset create an environment where committing mistakes is not viewed as a crime and where employees are fired but seen as part of the learning process, fostering resilience and innovation. Lastly, it is essential to empower employees through innovative learning opportunities. Manufacturing organisations must set up resources where employees are trained on leadership skills such as relationship building, adaptability, critical thinking, conflict management, and negotiation. Such workers help build employee confidence in their capabilities and contribute meaningfully to the organisation's innovative endeavours.

Practical Strategies to Develop an Innovation Culture within a Manufacturing Organisation

1. Foster a diverse and inclusive environment.

In the work environment, leaders need to be actively involved in fostering an environment where every person working for the organisation feels valued and appreciated despite their race or ethnicity. Fostering a culture of inclusivity ensures that employees do not have to work in the shadows as they are free to express their work capabilities (Graham, 2019). Also these employees feel comfortable enough to express their opinions and ideas. The power of diversity and inclusion must be considered as they lead to increased creativity. People from different backgrounds seem to have different ideas; thus, creating a team that is made up of people from diverse backgrounds means that you have brought different ideas into the organisation, leading to more innovative and creative solutions (Graham, 2019). Inclusive leadership enables individuals working under them to propose a more diverse range of concepts since they listen actively to the ideas and feedback from other people even when the proposed opinions do not support their ideas. Leaders who promote diversity also possess empathy by considering each employee's experiences and challenges as each staff member has different skills and capabilities, considering diverse perspectives and ensuring the decisions they make align with the needs and values of the team (Graham, 2019). Empathetic leaders also help organisations retain employees as they become more committed to their work.

2. Encourage intrapreneurship among employees.

Leaders in manufacturing organisations must promote intrapreneurship, where employees take full responsibility for creating their ideas or products once they're approved and implemented. It gives employees a reason to be dedicated to their work and committed to seeing the organisation succeed. By delegating responsibility to the junior staff, leaders allow them to learn about the company's operations. It helps increase job satisfaction levels among employees by offering them an experience that may not be possible in a typical working environment (Huang et al., 2021). According to research by *Forbes*, 40% of millennial staff were more interested in learning intrapreneurship and having an opportunity to pitch their ideas (Huang et al., 2021). In ensuring the effectiveness of intrapreneurship, leaders need to be precise on what the business is looking for, such as to create a new product or enhance the existing product and acknowledge that the proposed idea may fail; therefore, there is no need to blame the employee as it provides them with an opportunity to learn, ensuring they are not afraid to attempt again.

3. Invest in innovation and technology.

In meeting the ever-changing needs of consumers, leaders need to invest in innovative technology as it can proactively meet and exceed customer expectations. Using technology in manufacturing, a company can personalise goods depending on consumer preference, creating a better consumer experience and building lasting relationships with a client (White and Bruton, 2014). In addition, investing in recent technology enables an organisation to align its company objectives with sustainability goals. By using friendly technologies during production, the organisation can positively contribute to environmental conservation by minimising pollution such as land pollution through the dumping of used non-degradable products, air pollution through the emission of greenhouse gases, noise pollution from high sound generated from industrial equipment, and illegal dumping of contaminated water into river sources (White and Bruton, 2014). Investing in technologies such as using renewable energy such as solar, wind, or geothermal will help reduce the reliance on electricity generated by burning fossil fuels during production since renewable energy is cost-effective and accessible (White and Bruton, 2014). Advanced sensors such as data analytics and automation can be utilised to monitor and manage resource consumption, such as water in the production and cooling of machines. Organisations can contribute to a sustainable and responsible business model by utilizing different technologies.

4. Communicate long-term goals to staff members.

Communicating long-term goals is essential since it makes employees understand what is expected and helps minimise future risks. Open communication about long-term goals makes it possible for the staff to understand what the organisation expects of them in promoting sustainable growth, adaptability, and innovation (Craig, 2018). In the manufacturing industry, communicating long-term goals acts as a guide in promoting future research and development since they can anticipate future market trends and invest in emerging industry trends, which makes it possible for the organisation to stay ahead of its competitors and respond to market needs with innovative products before their competitors do (Craig, 2018). With the insights of long-term goals, the organisation's leaders would build strategic partnerships that help accelerate innovation. The manufacturing organisation can stay up-to-date with the latest research by entering into partnerships with different organisations, such as research institutions, private businesses, and technology companies (Craig, 2018). The research can be utilised in creating a new product or marketing their product, ensuring they can reach a wide range of consumers and strategically navigate uncertainties in the future.

Challenges in Promoting Innovation within the Manufacturing Organisation

1. Resistance from employees

Staff reluctance to implement new ideas or policies is common in many companies. Most employees fear change because innovation could make them lose their jobs due to the automation of work processes (Bilichenko et al., 2022). Additionally, staff may need to learn how innovation is essential for the company's growth because they are okay with the position they occupy in the organisation.

2. Increased workload and time constraints

When employees are allocated many tasks to complete quickly, it discourages innovation. Overworked employees need more time to engage and become innovative as there is no time to allocate tasks beyond their immediate responsibilities (Inegbedion et al., 2020). Overworked employees also suffer from fatigue and mental exhaustion, compromising their abilities to become creative and find viable solutions to problems within the organisation as this responsibility is left solely to the individuals in the management position (Inegbedion et al., 2020). Developing innovative solutions requires focus and the utilisation of the brain, which becomes a challenge since they are overwhelmed and suffer from burnout, lacking motivation and enthusiasm to explore new ideas or products.

3. Power structure within the organisation

Innovation may need to be improved in an organisation where power is in the hands of a few people within the leadership. This power structure allows the decision-making process to go faster. Still it does not empower the employees since the management leaders make most of the decisions (Kalay and Lynn, 2016). As such, employees often need more power to contribute their opinions, making it a challenge for ideas to flow among the staff in the organisation. It further slows the implementation process of ideas as they go through a lengthy approval process before staff can implement or start working on their ideas (Kalay and Lynn, 2016). Strict hierarchies interfere with effective communication between the employees and leaders, making it challenging for them to engage in creative problem-solving adventures and creating an environment where employees are discouraged from taking risks, experimenting, and being part of the innovation process.

Solutions for Combating the Challenges of Innovation within the Organisation

1. Implementing innovation processes and tools

Leaders can tackle the challenge of resistance among employees by utilising innovation tools to help educate employees on the importance of innovation within organisations. One of the tools that leaders can implement within the manufacturing cooperation is idea management systems, which provide a structured framework to facilitate and implement ideas from the staff. By utilising these systems, employees can work as a team to contribute their ideas by submitting them to a submission portal, and they can track the progress of their ideas in the system (Tams, 2020). Another tool that a leader can utilise is Crowd City, which is an idea management software that empowers the leader to create different challenges that the staff can respond to by offering their suggestions, allowing them to contribute to the work organisation while allowing the leader to be in control of the discussion board (Tams,

2020). Using this discussion board makes it possible to identify ideas from the staff that benefit the organisation.

2. Offering incentives to encourage innovation among employees

Leaders can utilise incentives to encourage innovation among employees within a manufacturing organisation. Leaders can utilise different incentives, such as financial incentives, work promotion, employee-employee recognition, and job security, to inspire and motivate them to create new ideas that enhance the organisation's image in the market, making it stand out among its competitors (Luo and Viki, 2020). When leaders offer incentives to their employees, they feel that the organisation values their hard work and are motivated to tackle even complex tasks. Offering incentives to employees encourages them to complete their tasks on time, leading to productivity as they are motivated to work with the possibility of reward (Luo and Viki, 2020). The utilisation of incentives inspires staff to identify problems within the organisation and devise viable solutions for addressing the identified problems, contributing to the overall efficiency and effectiveness of the organisation.

3. Encourage risk-taking among employees

Innovation involves employees taking the risk that their ideas will yield positive results. Thus, as a leader, it is essential to refrain from fostering a culture where employees fear implementing new ideas because they fear the consequences (Lendel et al., 2015). As such, it will deter employees from taking risks, resulting in a lack of experience and confidentiality; thus the people within the leadership positions in the organisation will spend most of their time supervising the staff (Lendel et al., 2015). Still a positive, safe environment where employees can see their ideas through to the end of the implementation processes empowers them to take on more responsibility and begin working independently, requiring little supervision as they feel happier with the project they have selected to handle.

4. Adopting a decentralised power approach

Adopting decentralisation of power within a manufacturing organisation allows the authority to focus on making critical decisions while delegating other tasks to managers at levels within the leadership hierarchy. This allows managers within the organisation to develop the staff's preparation and prepare them for future leadership positions (Kalay and Lynn, 2016). Further, when staff within an organisation feel that the people in leadership positions hear their opinions, they are more likely to commit to their work and be willing to work with others to exchange ideas, promote diversity, and fuel innovations across different departments.

Conclusion

In conclusion, adopting innovation within a manufacturing industry can enhance efficiency, promote collaboration, promote diversity, increase profits, and improve staff performance. Leaders within the organisation play a central role in leveraging innovation strategies to empower the staff by promoting independence. Workers who have demonstrated high levels of responsibility can work with minimum supervision and make decisions independently, promoting staff autonomy. This empowerment helps boost confidence and reinforce trust among the employees and their leaders. In addition, innovative leaders create an environment where staff members can take risks without fear since their leader allows them to learn from their mistakes and actively seek solutions to the errors they have committed. Thus, innovative leaders foster a culture of innovation and put the company in a position to compete effectively globally. Lastly, with emerging manufacturing industries changing, leaders must prioritise innovation strategies to help meet consumer needs, navigate market challenges, and prepare them for a better future.

References

Aslam, H., Waseem, M., Muneeb, D., Ali, Z., Roubaud, D., and Grebinevych, O. (2023) 'Customer integration in the supply chain: the role of market orientation and supply chain strategy in the age of digital revolution', *Annals of Operations Research*, pp. 1–25. https://doi.org/10.1007/s10479-023-05191-y.

Bilichenko, O., Tolmachev, M., Polozova, T., Aniskevych, D., and Mohammad, A. L. A. K. (2022) 'Managing strategic changes in personnel resistance to open innovation in companies', *Journal of Open Innovation: Technology, Market, and Complexity*, 8(3), p. 151. https://doi.org/10.3390/joitmc8030151.

Craig, W. (2018, August 14). 'Why communicating company goals is key for employee growth', *Forbes*. https://www.forbes.com/sites/williamcraig/2018/08/14/why-communicating-company-goals-is-key-for-employee-growth/?sh=50d9bfc13b85.

Gharam, J. (2019) 'Enhancing diversity: The manufacturing mindset advantage', *Insights*. Heidrick & Struggles. https://www.heidrick.com/en/insights/diversity-inclusion/enhancing_diversity_the_manufacturing_mind_set_advantage.

Huang, L. Y., Yang Lin, S. M., and Hsieh, Y. J. (2021) 'Cultivation of intrapreneurship: a framework and challenges', *Frontiers in Psychology*, 12, pp. 731–990. https://doi.org/10.3389/fpsyg.2021.731990.

Inegbedion, H., Inegbedion, E., Peter, A., and Harry, L. (2020) 'Perception of workload balance and employee job satisfaction in work organizations', *Heliyon*, 6(1). https://doi.org/10.1016/j.heliyon.2020.e03160.

Lendel, V., Hittmár, Š. and Siantová, E. (2015) 'Management of innovation processes in the company', *Procedia Economics and Finance*, 23, pp. 861–866. https://doi.org/10.1016/S2212-5671 (15)00382-2.

Luo, L., and Viki, T. (2020, July 23) 'Innovation culture: Rewards and Incentives', *Innovation Culture: Rewards and Incentives*. https://www.strategyzer.com/library/why-its-important-to-give-innovators-a-stake-in-innovation-success.

Mitcheltree, C. M. (2021) 'Enhancing innovation speed through trust: a case study on reframing employee defensive routines', *Journal of Innovation and Entrepreneurship*, 10(1), 1–31. https://doi.org/10.1186/s13731-020-00143-3

Tams, C. (2020, May 13) 'Getting ready for the ramp-up: The Power of Idea Management Platforms', *Forbes*. https://www.forbes.com/sites/carstentams/2020/05/06/getting-ready-for-the-ramp-up-the-power-of-idea-management-platforms/?sh=2fdc350448a3.

White, M. A., and Bruton, G. D. (2014) *Strategic Management of Technology and Innovation*. Cengage Learning Asia Pte Ltd.

Top-Down Innovation (TDI) versus Down-Top Innovation (DTI)

I have been practising various methodologies and innovation behaviours to create an innovative environment amongst the employees and teams in the organisation. Let me explain a simple concept called top-down versus down-top Innovation.

Figure 4: Top-Down versus Down-Top Innovation

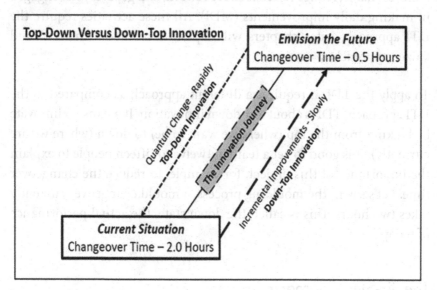

Source: Azlan Nithia (2023)

I have practised both of these concepts for many years. The concept of down-top innovation (DTI) is very commonly practised in organisations. This is the practice that organisations refer to as a continuous improvement program simply called the kaizen way.

The DTI way is about making small improvements, now better than yesterday or this month better than last month. In this way, we make incremental improvements, step by step, getting better and better over time.

In times of extreme manufacturing crises, which method is better, applying TDI or DTI? In crises, the number one priority is to get out of a crisis, nothing is more important than that. In crises apply, down-top innovation, improve one step at a time, continue making improvements every day, and make this week a better week than last week and this month better than last month. For example, focus on 5S in every area of the organisation, stop unsafe acts, improve unsafe conditions, reduce the machine downtime, improve the manufacturing processes to make it more reliable, reduce rejects, reduce rework, and get everyone engaged in making daily improvements (MDI). All these activities require the DTI approach. Later chapters will explain more about the lean and kaizen approaches.

To apply the TDI, it requires a different approach as compared to the DTI approach. TDI is about top-down innovation. It means we innovate by looking from the top (where we want to be) to down (where we are currently). It is good to get a team of twelve to fifteen people to explain the importance of this project, for example, to reduce the changeover time. Let's say in the moulding process, a mould changeover currently takes two hours (this is called the down state, the actual performance of today).

The Six-Step Approach

Firstly in step 1, the team must be convinced and agree on the importance of the critical benefits the company will have or gain when the top results are achieved. The top result (envisioning the future results) is an important point, to envision looking from the top. In the mould changeover time, the top is 0.5 hours (30 minutes), and the down is two hours (120 minutes). The difference between the top and down is ninety (90) minutes. We need to reduce 75% of the changeover time. It may look almost impossible, but 75% is a quantum leap to achieve the 0.5 hours.

Next step 2 is to break up the team (12 to 15 people) into two or three smaller teams. Think of the future (top) the 0.5-hour mould changeover time, the process, the activities, and who is doing what. Close your eyes and visualise a thirty-minute mould changeover process (from start to finish) is actually happening. What does it look like, what are the activities that are happening in the process of changeover and making the 30-minute possible? Next as a team, in your individual teams, list down all those activities that are making the future possible. All three teams work separately.

Step 3, get all three teams together to present to everyone, what they envisioned the future (thirty-minute mould changeover) process and activities will look like. The teams share the ideas, capture the best ideas from each other, and then agree on the final process of the future that makes a 30-minute mould changeover process possible.

Step 4, change the team members and recreate two teams from the current members. The two teams work separately to chart out how to achieve the 30-minute changeover process, what needs to happen, who needs to do what, what obstacles need to be overcome, and how to overcome them and help required, financial investment (if required), potential savings, and the timeline to make it happen.

In step 5, both teams get back together. Combine both the team's actions into one. Agree on the final execution plan and the timelines, all agreeing to only one better execution plan. Included in the execution plan will be the actions that can potentially derail this project (what can potentially go wrong) and the actions to prevent those potential derailments from happening.

Step 6, make it happen by the execution team, implement the execution plan, create success in one small section first, gain confidence, and celebrate the small win. Next is to expand the implementation scope further till the full completion of the project is done.

The execution team must meet weekly to give the leadership team the weekly update, progress, and results achieved so far and for the leaders to provide any support needed by the execution team.

The above six-step approach is to achieve or implement the top-down innovation (TDI) that demonstrates how to achieve the 'quantum change' that is implemented rapidly.

The TDI approach is recommended when the organisation has gotten out of crisis. The organisation is stabilised and when employees are ready to compete. This is when the TDI approach will drive (or deliver) bigger competitive successes and create greater growth and better profitability. The champions (the employees) will be delivering the winning results, gaining a high level of confidence, and marching towards becoming a globally competitive organisation.

Building a Champion Team

The high-performing teams or what I like to call 'the champion team' are the champions who will relentlessly commit to changing what is not good, make it better and even better, very passionate, and work tirelessly. They rely on a blend of actions in order to achieve what they do and focus on the required results to deliver. While much of the management focus is understandably spent on trying to manage individuals, personalities, and the ways to motivate them, success cannot be achieved if the right skill sets are not trained or hired in the first place and then introduced in a way that builds easy cohesion.

It is always good to train internally and promote internally. Learning the high-performing internal culture for a new person takes a long time and can be complicated for someone newly joining the organisation. Hiring from outside the organisation must be the last option. This will be done if the organisation does not have anyone with the required skills. A well-planned succession planning is extremely crucial for any organisation or function. The person being hired from another well-run organisation may not have the experience to manage a function in crisis or to revive a broken function. To continue managing a well-run function and to continue making improvements are not complicated. It goes on a balanced autopilot but is not an organisation in crisis. Many individuals, who are hired from a very successful and well-ran function, have failed in another organisation that is in crisis.

We need vehicles that are built for rough roads to efficiently function on the rough roads successfully. Comfortable cars are not built for rough terrains. They could be broken before we even complete the required ride distance!

I have always found that identifying the rising talents internally is the best. Training them on the required technical and leadership skills over a period on a structured timeline is important. Allow the rising talent to work on an expanded job responsibility over a period of six to twelve months till results are achieved, and they are very evident. Identify a good mentor for the rising talent to assist and ensure success in the various challenging roles.

It is important for any organisation to have a talent development pipeline. Based on my many years of experience, I have found that having a good internship program is a good source of future talent pipeline. Interns for this internship pipeline purpose must be selected through a rigorous selection process by recruiting the best and most capable young talents. These interns must be trained in all lean methodologies and kaizen improvement activities, After six to twelve months, we should have a pool of young and very energetic talents ready to take on functional roles, putting them on some problematic jobs to make improvements by assisting the current functional supervisors and area functional leaders and allowing them to make improvements and progress by doing 'one-step-at-a-time' approach. They must approach difficult situations with systematic problem-solving skills (which will be discussed later in the book), implement effective solutions, show results, and not excuse givers but be a solution provider.

Create a talent development pipeline—a good internship program! We need good problem solvers and solution providers!

The leader of the organisation must always find time to speak to those key rising talents and allocate at least thirty minutes every week for one-on-one sessions with all those key employees. If the leader does not allocate time or he/she is too busy on other things but not this, then there is a serious problem of leadership. The leader failed to understand his/her key role in developing a team of champions.

Champions learn by doing the right things and the right way very rapidly. Therefore, the champions team is also a team that learns new things very fast, deploys critical thinking, and engages all their team members in the direction of the new better way to achieve the company's strategic results.

Have the right people. Do the right things and the right ways.

As we know when our people move on and get promoted, they will need to be replaced in the most efficient way, this could be a frequent process at all different levels of the organisation.

It's no wonder that it is an ongoing journey for every company, and it may get more complicated as we move towards a more digital or virtual world.

Succeeding on this journey requires a clear understanding of the competencies needed for each role, not only today but in the future as well as knowing how to check that these exist during candidate interview.

While competency and experience are important, so are attitude and desire to do the job required. Many a hire can fail because the interview hasn't covered what is needed in these areas in the depth required.

The journey also requires clear communication across the team, especially during the onboarding process. It also requires the ability to address any performance or expectation issues as and when they appear early on rather than wait until performance review season.

Hiring is expensive and failing to hire the right people will only further add to the cost in a much bigger way.

How to Retain Champions at the Workplace

The biggest challenge for any organisation is to retain good people and that they will always be our high-performing champions in our organisation. How do we retain these employees? The freshers when they come in, within six months, one year after you have trained, and they leave. So that is a big challenge which we are still trying to address. We are thinking of looking at some kind of a retainership program for at least three years. This will not work! I don't know, so I am asking! I am trying to figure out how to do this.

I want to share this—the world of business has seen three generations. The Industrial Revolution, the information revolution, and now the social revolution.

Industrial Revolution—people took jobs for survival. They wanted basic necessities of food to eat. So they joined in, they worked in a workplace, and they did not leave that workplace even if the boss was very abusive and even if the boss would physically beat them up. However, people would not leave. That was the era of 'boss is always right'. This was because the opportunities were less. That was pretty much the grandparents of today's workforce.

Then came the **information revolution**—where all these IT companies and brands started building, and the workforce then came to work not for survival because their parents took care of their survival. This workforce went to work for the standard of living. Basically get a good enough salary to pay the house loans, car loans, and children's education loans. It was pretty much our generation. We went to work for that for the standard of living. In this generation, loyalty is reduced. The truth is that loyalty never existed in the first place. Loyalty never existed. There was no option. This generation, the information revolution has options. If they got a better quality of workplace and they got better pay, then the people will jump a job. This was all the information revolution.

Then on the third revolution after 2008's recession, the information revolution died. Today information is available for free. You can learn coding on YouTube. You don't have to pay to learn. It's no longer a hidden knowledge.

So we need to understand. Now we live in a **digital revolution or a social revolution** where everything is about social, which means today's workforce—they don't care about survival. Their grandparents took care of that. Even the labourers do not care about the standard of living anymore because even labourers dish TV and Tata Sky in their houses. So if you say I will cut your pay or if you say I will bind you with a contract, they will say, 'Thank you so much, I am finding another job.'

Today's workforce needs something else. They want quality of life, not standards of life. Their parents took care of the standards of life. Today's workforce is thinking about the quality of life, which means the quality of the workplace which is the quality of job, quality of the environment, quality of role, role, opportunities, learning, and rewards—all of this. Unless you don't have a mechanism for that, you will always grapple with this retention challenge.

CHAPTER 5

Lean and Leadership

Senior Leaders Commitment

To begin the lean journey and create a lean culture is hard; it needs the organisation's leadership team's commitment for the long run and a long-term perspective. The journey is at all times ongoing without a finish line or an end date. It is better not to start this journey if the senior leaders and the management team are not willing to begin and remain on this difficult but important journey. This journey is part of a larger movement, which is the enabler of manufacturing excellence required to establish a strong foundation for 'smart manufacturing', the essential requirement for the 'factory of the future' and to achieve the vision of a 'connected factory'. This must be realised with increased customer responsiveness, low cost, and mass customisation (small lot sizes, mixed model production, quick response, and increased degree of product variety). New organisational work cultures are not created in a few months as it takes years coupled with a passionate commitment to operationalise a seamless new work culture for it to be embedded in the company's DNA. Achieving this requires genuine senior leadership involvement as this is the key to the success of this lean journey. Without it, the effort is limited, and the operations team is powerless.

The top leaders must be committed to building the kind of management in which everyone can contribute directly to adding value for customers and take responsibility for leading the culture change. Failing to see this as a major required cultural change and assigning the program to middle managers to initiate is a major mistake. A business transformation that puts customers first and does it by developing its people has to start with the top leader's commitment. Lean is not only about removing waste even though this is the key component. Being lean is about creating a new culture and living it.

The leadership cannot deploy a lean structure, calling it a lean office, employ continuous improvement officers, and assign them to apply this lean culture in every area of the organisation if the management continues with what it is used to do by not being fully involved and not staying passionately interested. If this situation ensures the management team will be surprised that after gathering all the low-hanging, easy-to-do improvements, the whole lean culture sinks. After a while, the lean effort is abandoned, and the management looks for the next program of the year to embark on. The lean journey and creating a culture of continuous improvement is not a toolkit or a road map. Everyone in the organisation, top-down, must understand the lean effort, live it, and continuously evolve it.

There are certain organisations where the senior management leadership, CEOs, and CFOs view the manufacturing operations as a division that derails the focus of running the business, which is very capital intensive. Handling people's issues is difficult. It is a dirty and difficult job. They rather have some other company do the manufacturing portion. This view is a clear indication of the senior leadership lacking the understanding and knowledge of how and where the product value is created. The value is created when the organisation has total control over the supply chain, total lead time, enhanced quality, reduced cost, and customer responsiveness.

In my opinion, organisations that already have their own brands (or IPs), in-house manufacturing, product design and development capabilities, own marketing, and sales teams must seriously commit and stay aligned on the lean journey. They have total control of the supply chain (end to end), and the manufacturing function is an excellent value creator to the business both upstream and downstream. Some of these examples are companies like the car industries (Honda, Toyota, Mercedes, Nissan, Ford, etc), Lego toys, Samsung, Robert Borsch, B Braun, Nestle, and many more successful industries. These are the organisations that have total end-to-end business control from product creation and development all the way to delivering the products to the consumers with costs the consumers are willing to pay. These companies have total control of their supply chain coupled with the ability to continuously improve performance by designing products for lean manufacturing, total lead time reduction, inventory control, quality, total cost, and continuously improving customer responsiveness.

There are also certain organisations that already have total control of the end-to-end supply chain with in-house capabilities for product design (own brands), manufacturing, marketing, and sales. Sadly some of these organisations encounter strategic changes when new senior leadership, CEOs, and CFOs included and have a very different perspective about the current manufacturing operations. They may have no manufacturing experience and an understanding of how the manufacturing division creates value and constantly adds value to the business. Instead they view manufacturing as a difficult burden to manage, very capital intensive, and bad for their cash flow. Subsequently the organisation changes its strategy. They rather buy versus make it in-house (manufacturing). These are the senior leaders not imbued with the understanding of the power of lean strategy or lean manufacturing and how manufacturing plays a key role in improving customer responsiveness, reducing lead time, and improving cost. This kind of poor short-term thinking management perspective will derail the organisation's end-to-end future capabilities, especially when these senior leaders make decisions to shut down their existing manufacturing divisions.

When taking control of the total end-to-end supply chain and delivering products at shorter lead times and lower costs, the organisation should implement the design for lean manufacturing to achieve high-manufacturing productivity and efficiencies. This delivers the organisation a massive advantage and opportunity to achieve the best cost and lead time. If the organisation outsources the manufacturing, having a 'buying' instead of 'making' strategy, it will be very difficult for the organisation to move forward with the initiatives to continuously improve or reduce lead time and cost. It will become dependent on the outsourced vendor's manufacturing strategy and capabilities coupled with their manufacturing systems limitations.

To successfully begin this lean journey to achieve manufacturing excellence and later embark on smart manufacturing, it will solely depend on the senior leader's commitment during this difficult but important journey in creating a strong foundation. Lean is about developing the kaizen spirit (continuous improvement, continuous learning) in every employee.

Three Drivers of Manufacturing Competitiveness

Achieving this manufacturing excellence is a journey that must be pursued by both the organisation's leaders and employees. This can be realised by actively engaging and cultivating a culture of continuous improvement. The dedicated focus must be to improve three important high-level performance drivers of organisational competitiveness in the marketplace in order to achieve manufacturing excellence.

The three manufacturing excellence drivers are

1. **Lead time reduction**

To continuously reduce the total end-to-end supply chain lead time and model changeover time reduction, focus on agility and eliminate any process rigidity. To always meet the customer delivery commitments of on-time, in-full deliveries (or OTIF). There are many components of activities and events in the manufacturing systems that contribute towards the lead time, examples of which are repeated product design changes, rework, waiting for a decision, supplier's lead time, and many more.

Lead time is a key driver of competitive advantage in the marketplace and is critical to the customers. It is also a key driver of cost in the organisation. Therefore, to reduce lead time is to reduce or eliminate all the obstacles in the flow of the products and information. Lead time reduction is not a one-time event; lead time reduction must carried out constantly, year on year. The method used to establish the current lead time is by doing the VSM (value stream mapping). From the VSM kaizen, the VA (value added) and NVA (non-value added) work and activities are identifiable. The NVA must be the key focus. The company must constantly embark on reducing the NVA, which also means lowering the constraints and obstacles to the flow. By reducing these obstacles, the flow increases; the increase in process and product flow reduces lead time and increases throughput. It can be equated to a river. When there are many rocks and blockages in the river, the flow of water in the river is not smooth, and the flow is frequently obstructed. When all the rocks and blockages are removed, the river flows very smoothly. One can see a smooth flow of water in the river without any turbulence. This analogy applies to any manufacturing operation in any industry.

The organisation must complete a VSM kaizen every year and constantly keep identifying and removing the NVA from the system, and these actions required to reduce the NVA can be accomplished in a kaizen event. The VSM activity will be covered in the later chapters. Reducing or eliminating inventories (inventories are also considered as an obstruction to flow), and the work in process (WIP) is one of the key objectives of the VSM kaizen. The inventories are the obstacles to the flow (like rocks in the river), and they reduce the speed of flow. Therefore, the reduction in the speed of flow will increase the lead time. It is very important to understand the concept flow in manufacturing. The best flow is achieved only when there are no obstacles to flow. The *muda*, *mura*, and *muri* (3M) waste creates obstacles to flow. By reducing these obstacles, the flow increases, the lead time reduces, and the throughput rises. When the lead time is reduced, this will afford more available time for production, thereby increasing the throughput.

2. Product cost reduction

The organisation must continuously reduce the total cost of the product and the cost of services for the customers, year on year. Always focus efforts to reduce and eliminate the 3M waste in the system, as all of them impact the cost negatively, by constantly striving to reduce them and reduce the product cost. The ultimate concept of cost must be that expenditures in the present year must be lower than the previous year without sacrificing quality or moderating customer satisfaction.

There are numerous economic elements that will impact the product cost, examples of which are material cost increases, salary increases, and other services-related cost increases. An organisation must always strive to mitigate all these cost increases by focusing on various cost-reducing programs, especially the 3M waste. The organisation must never attempt to pass the cost increases to the customers because in the long run, there will be other competitors who will be able to deliver better cost and lead time. If the organisations are always internally focused on reducing operational and material costs and improving quality, then mitigating any cost increases becomes very possible. Organisations that do not focus on year-on-year product cost reduction will soon find themselves eliminated from marketing competition.

3. Improving the throughput (productivity)

The throughput in most cases will increase if the lead time or the process time is reduced. To continuously improve the production throughput means producing more output with good quality products (that converge on customer expectations), meeting all the customer requirements, and always fulfilling customer delivery expectations (on time, in full). The process engineers and supervisors must understand the lean flow and the constraints and identify bottlenecks in the flow. The implementation of a strategy to discard 3M wastes is critical, constantly finding the mura and muri in the system and smoothening out the constraints and the flow obstacles. When the bottleneck process constraints are improved, the output increases. It is a relentless effort to continuously focus on finding, reducing, and eliminating constraints (or the processes that are bottlenecks) in the system. It is usually common to keep finding new constraints after the current one is resolved or after the earlier constraints have been eliminated. It is an ongoing journey of continuously finding and resolving the process constraints, working towards achieving perfection in manufacturing flow and continuing to optimise output and productivity.

The two charts below show the lead time reduction (new product launch) and product cost reduction, both based on an actual case study (done over 10 years) and the achievement of the organisation that was very focused on relentlessly reducing the lead time, increasing throughput (or the output) and the cost, year after year, continuously improving in those three important KPIs.

After ten years of continuously improving (reducing) the lead time and the product cost, the results of business competitiveness are obvious. The lead time was reduced from 40 weeks to only 15 weeks (270% reduction), and the product cost was reduced from $4.50 per item to only $1.90 per item (230% reduction). With the constant reduction of lead time (relentlessly reducing the 3M waste and improving the flow) the total throughput of the company was increased from five (5.0) million units a year to thirty (30) million units a year, sixfold throughput increase and with lesser resources. This was achieved even with the constant rise of cost pressures, inflations, increased customer expectations and complexity, increased material cost and mitigating the yearly salary increases.

Figure 5: Lead Time Reduction

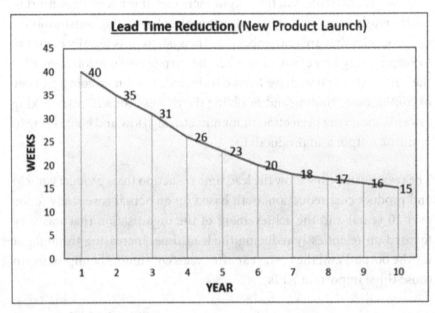

Source: Azlan Nithia, 2019

This company is now well-positioned to compete globally in its business domain, eliminate its competitors, and be a global leader in the industry. It is about competing against its own performance. The company was never satisfied with its current performance. This is the lean spirit of driving continuous improvement behaviour and living it daily. Currently this company is extremely competitive in cost, quality, and lead time. It has been almost impossible to find another company to outclass them.

Never stop and continue improving!

Figure 6: Product Cost Reduction

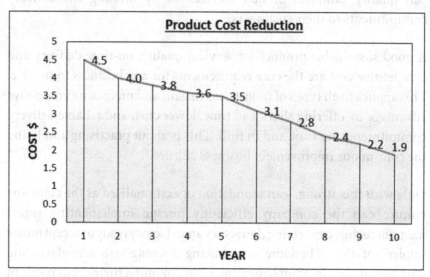

Source: Azlan Nithia, 2019

There are not many organisations in the world today that have aggressively pursued these kinds of relentless lean journeys. Toyota's TPS (Toyota Production System) is an excellent example for the world to emulate, the lean journey that Toyota started in the 1950s. It is an embedded DNA in Toyota's leadership and employees all around the world.

Those strong lean foundations and organisational culture that were implemented and operationalised in the above company have developed an efficient smart ecosystem in this company to implement the smart connected factory of the future. After completing the implementation of the connected factory and the IoT projects, this company has further improved its customer responsiveness, further reduced lead time and cost, and better product quality.

The continuous improvements in lead time, product cost, and throughput (productivity) must be compounded with good quality and OTIF (on-time, in-full) delivery. The quality and delivery commitments are the basic requirements for any organisation to stay in business. An organisation cannot survive in any business or industry by delivering bad quality products or bad services or by missing the delivery commitments to their customers.

A good sustainable product (or service) quality, on-time delivery and competitive cost are the core requirements for any business to survive. This applies to all types of industries. To gain and maintain competitive advantage by offering short lead time, lower cost, and reliable delivery commitments (on time and in full). This is about practising and living the continuous improvement business culture.

Only with this strong, lean foundation operationalised as the company culture can the company efficiently pursue implementing smart manufacturing, connecting the factory and a factory ready to accept future implementation. Therefore, only having a strong lean manufacturing foundation and then implementing smart manufacturing will result in improved operational efficiencies, and the factory of the future will then be able to deliver better customer responsiveness, reduced lead time, and lower cost. Directly implementing smart manufacturing without a strong lean foundational culture can be disastrous for the operations.

Linking poor-performing machines with IoT (internet of things) to a centralised monitoring system will not provide better output nor will improve the performance of the machines. The foundation is to have a good-performing machine, this being step 1, predictable and capable machines to consistently deliver good parts. Then the next step is connecting these machines with IoT and centralising a decision system that will deliver good value for the investment (ROI). This will improve responsiveness and quality and reduce cost.

The 3M Wastes in Manufacturing

The 3M wastes are very often overlooked by most organisations, and these wastes exist in every organisation and industry.

There are three different types of waste in any organisation or industry. Most organisations embarking on the lean journey focus only on one type of waste, which is commonly called the *muda* in Japanese, or simply called the waste. To achieve manufacturing excellence there are two key drivers of waste in the manufacturing system, they are called the *mura* (unevenness in the system) and the other is called *muri* (overworked or overburdened process or person).

Omitting *mura* and *muri* is a big mistake we make in any manufacturing operation or industry because most of the waste (*muda*) is created by *mura* and *muri*. These 3M wastes are the important focus of the Toyota Production System (or TPS).

These three types of waste *muda*, *mura*, and *muri* are the 'enemies' of lean and must be reduced and ultimately eliminated to achieve manufacturing excellence.

- *Muda*—**waste**, means activity or process that does not add value to the customer, a physical waste of time, resources, and material
- *Mura*—**unevenness**, means waste of unevenness or inconsistency of processes and loading unevenness
- *Muri*—**overburdened**, means that operators or machines are pushed through their natural limits, which leads to problems.

If the organisation does not seriously embark on the relentless journey to reduce and eliminate the 3M wastes, then achieving manufacturing excellence cannot be assumed to be conceivable.

'Eliminate 3M waste = Manufacturing exellence'

It is important to understand how they are interrelated and influence each other. Mura creates waste (muda), and muri creates waste (muda). Together they impede manufacturing excellence.

Therefore, reducing mura, muri, and muda must be studied simultaneously as a never-ending quest to eliminate the 3M waste that will enable manufacturing excellence. This must become the organisation's relentless journey to pursue the continuous improvement culture which forms one of the important pillars of TPS.

Muda (Waste)

Muda is any work that is done by the organisation that the customer is not willing to pay for, does not add any extra value to the product, or is an obstacle to the process or production flow. This non-value-added work or obstacle (muda) to flow normally leads to an increase in lead time, increases product cost, and impedes throughput performance.

Muda is any activity or process (in manufacturing or business process) that does not add value to the customer, a physical waste, resources and material waste. This is the activity carried out by the company but does not transform the product of which a customer does not see or care.

There are eight types of waste (*muda*) in manufacturing:

1. Defects

- These are the products or services created with a defect and moved to the next process or the next step of manufacture. These products do not meet or conform to the customer's specifications or requirements. When these defects or defective products are allowed to flow to the next process, these defects create extra work to rectify or rework in order to uphold the quality of the product. This defines non-value-added work or activity. It interrupts the current work in the process, and when these defective parts flow to the next process or steps, it also interrupts the flow and rhythm of work done.

It is important to cultivate a work culture of pride.

'Make good parts, send good parts, and receive good parts.'

2. Overproduction

- Overproduction means producing more products than the customer needs at the current point of time. This could also mean the products produced are more than what the subsequent process requires. This creates unwanted inventories. Inventories take up space, hold up cash (parts produced are money already spent), and risk order cancellations. Practice just in time (JIT). Deliver the right quantity of products at the right time to the next process, nothing more and nothing less. It is important to measure on-time, in-full (OTIF) performance for the internal processes.
- There are internal and external customers, the internal customer being the subsequent or the next process in the system while the actual external customer is the final customer who receives the product or services and pays for the product or services.

3. Waiting

Waiting is when parts are waiting in the WIP (work in progress) on the shop floor, waiting for a decision, waiting due to pending paperwork, or parts are on hold due to poor quality. These are examples of the waste created due to waiting. Parts produced but are 'waiting' will interrupt the flow sequence of the next process. The company must establish a system that triggers immediate action. A machine breakdown results in parts that cannot be produced because the process is 'waiting' for a technician to repair the machine, an example of waiting. This waiting due to a machine problem will create part shortages and delivery delays. This will affect the part OTIF performance. The company must institute an Andon (visual system) that triggers a technician to take immediate action. It is important to measure the waiting time and continuously focus on reducing this non-value-adding waste. As for the machine breakdown, it is recommended to measure machine or equipment breakdown frequency and the mean time between breakdowns.

4. Transport

Transport is the movement of products between workstations and processes involving multiple individuals. Any kind of movement is a waste; therefore, it is critical to create a part movement path or route that takes the shortest time and involves the least resources. There are numerous methods this can be achieved. Examples would be by integrating various processes in the system, such as 'a flexible mini production line' that is linked into one system flow or by bringing processes in the system as close as possible.

5. Inventories

Inventories are products or services that are waiting for the next operation at a workstation and in the inventory holding stores or locations. Inventories that are waiting are considered waste. It is important that all inventories in the shop floor are accurately controlled by the kanban system and there should be a visual kanban system. The *kanban* system manages visual inventories, the minimum and maximum inventories, and the timely part replenishment orders system deployed. This system will ensure the right parts with the right quantities are in the inventory; nothing more and nothing less are available for the next process on time. Inventories are waste. Therefore, there must be a continuous effort undertaken to reduce them. Every part in the inventory must be prioritised with the FIFO system to ensure a continuous flow and an active customer requirement.

6. Motion

The movement of people and machines not integrated into the production or service is a non-value-adding or wasteful activity. Every movement of people on the shop floor must be reviewed critically, and finding ways to reduce those movements is a priority. Mapping out the movement diagram also called the 'spaghetti mapping' of the people is important. By mapping out the movement of the day or the shift, it is possible to find ideas to reduce those movements or even eliminate them completely. You will be surprised how many meters or even kilometres a person walks on the shop floor in a shift or a day. There are cases of employees walking up to five to six kilometres a day on a shop floor.

Similarly it is also possible to map out the movement path of a machine or a robot. There will be numerous movements or paths taken by the machine or robots that do not add value to the product nor change the product form when performing the actions. A robot that makes fanciful movements may look very interesting, but it does not add value to the product. It is imperative to delete the non-value-adding robot movements.

7. Excess processing

Excess processing is performing extra operations on the product that the customer is not asking for or seeing. These are wasteful activities. The quality function in the organisation must clearly understand the customer's expectations and establish quality control systems in the production processes accordingly so that the production activities are not carried out beyond the customer's expectations. There are numerous cases whereby excess work is carried out in the production by the employees on the product or in the processing due to lack of SOP (standard operating procedure) and lack of understanding of the customer expectations. An SOP must be established for every process in the system and all the employees must be well trained and certified to operationalise the SOP effectively all the time. Any deviation from the process SOP must be reviewed very seriously and immediate action must be taken by the process supervisor. The employees must not be allowed to deviate from the SOPs without supervisor or management approval and must become an organisation culture.

8. Non-used talent

Employee knowledge and skills, which are not utilised to their fullest potential in order to improve the processes, are regarded as wasted employee talent. As we have understood from the previous chapters, employee knowledge, involvement, and improvements are very crucial to achieving mass customisation and manufacturing excellence.

All improvement activities require employee involvement. Optimising every employee's talent in an organisation is important to develop and achieve manufacturing excellence and achieving manufacturing excellence facilitates the mass customisation process.

All the waste or *muda* can be classified as 'islands of waste' in the company. There are eight islands of waste. It is important to look at the whole system and then identify individual types of waste in the system. The process or methodology that is effectively used in lean, to systematically establish waste (non-value-added activities) in the whole system in an organisation is called 'value stream mapping' (or VSM in short). It is important to understand the kind of waste that exists in the system by completing a VSM kaizen (from start-to-end process mapping).

Mura (Unevenness)

Mura is the unevenness that is found in various stages of the manufacturing system. It creates constraints and loss of system throughput.

The examples of *mura* or unevenness are as follows:

1. Fluctuation and changes in customer orders
2. Variations in process times and variations of cycle times for different operators
3. Failure of JIT (just-in-time) or OTIF (on-time, in-full) delivery between the processes in the system
4. Failure of the 'takt time' process. Takt time is the cycle time rhythm for the whole system throughput. Any process in the system not aligned with the takt time will create unevenness and lack of uniformity in the process and output. This failure will impact the flow in the system and result in mura.

If the *mura* is not reduced or eliminated, it increases the possibility of overburdening (*muri*) the affected underperforming process, which will create *muda* (waste) by interrupting the flow through the system. This mura will impact lead time, outputs, and obviously the product cost. Mura creates most of the eight elements of *muda* (waste); therefore, most often *mura* creates *muda* and *muri*.

It is important to constantly identify and reduce unevenness, variations, and constraints in the production or manufacturing system by aligning all the processes.

Production Schedule Smoothening

The chart below depicts an example of poor production scheduling. The production scheduling method varies from one organisation to another, and the production scheduling is planned and executed based on monthly, weekly, or daily production and shipping targets.

The chart illustrates an organisational production schedule planned on a monthly production of delivery and shipping targets. The monthly customer's delivery target is 120,000 units of products. The normal production behaviour will be the last week of the month rush to meet the total order by the month-end cut-off date. This kind of production behaviour creates major unevenness in production processes and the system, especially in the last week of the month.

An organisation that operates on a weekly scheduling target tends to have an irrational rush to meet the targets on Friday and Saturday, creating unevenness between processes and inadvertently creating various kinds of bottlenecks on the last day of the week. This then creates major issues in throughput on Saturdays to meet the weekly target by Saturday night's shift.

Figure 7: Weekly Schedule—Unevenness (Mura Waste)

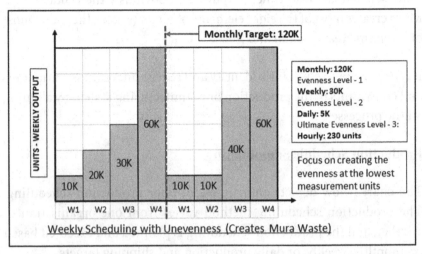

Source: Azlan Nithia (2023)

Then is daily target scheduling any better? Yes, daily target scheduling is much better than a monthly or weekly scheduling system. The daily production planning schedule assumes the production is operated in two-shift operations, working ten hours per shift, a morning shift and an evening shift. The target cut-off will be on the second (evening) shift, on the last hour, or the twentieth hour of the day, where the target must be achieved. If this is further analysed hourly, we will experience the major production loading challenges that would have been on the second shift, especially the last few hours of the second shift to meet the day's full target quantity. Therefore, the constraints will be created in the last few hours of the production on the evening shift. Again there is unevenness in throughput if compared from the first hour production output up to the last few hours of the second shift output.

The better solution to manage the schedule with the least *mura* or waste is by ensuing constraints are minimised will be to schedule production targets by the hour. Every process, machine, and system must be scheduled to deliver the required hourly output targets. Any interruption must be resolved within the same hour (ultimately the aim is to solve problems within minutes) to ensure a continuous flow of products through all the processes in the whole production system. In an hourly planning and scheduling strategy, the 120,000 units target will be broken down to an hourly target of 230 units per hour (operating 26 days a month and 20 hours a day). See figure 7.

Smoothening the production schedule to hourly segments is extremely critical to ensure process constraints are not insidiously developed anywhere in the whole system. Focus on operating on evenness must be executed even at the lowest scheduling units. Every process and equipment used must be tracked to deliver this hourly target, and every shift performance must be discussed by the production supervisor and actions taken to resolve it so that these issues do not continue affecting the next later production shift performance.

The front-line leaders (sometimes called line leaders) and supervisors must be actively engaged in the activities being performed at the *gemba* (workplace) so that immediate support and help can be provided to the workers having any problems with the process or machines. To facilitate this, workers calling for help or having a problem can be done using the Andon (visual trigger for help system). The Andon is triggered by pushing a button that lights up a red light and siren or by pulling a string that triggers the siren and light. The light must be visible and must be easily seen from afar by the technicians and supervisors.

A Case Study

Overcoming Constraints in the Production System

The figure below (constraints in the system) is an actual case study done on constraints in a production line. The poor throughput performance was affecting the customer order deliveries. These constraints impacted the overall system throughput. The throughput is always constrained by the worst-performing process; the best output is constrained by the worst-limiting process. There are a total of 35 stations (or processes) in this production line. The cycle time of the 35 processes ranges from 5 to 24 seconds. The longest cycle time processes choke the flow. In this case, it is 24 seconds (secs), stations number 1, 4, 25, and 27. Even though there are processes that are running at 5 secs, the throughput cannot be increased due to the processes operating at 24 secs. These are the *mura* in the system that creates unevenness in the flow and output.

The engineers did a very detailed study of every activity being done in those four limiting stations, namely 1, 4, 25, and 27. The station's operation sequence and activity were improved by 5 to 6 secs, reducing the mura or constraints to the throughput. After those constraints were solved, the flow was greatly improved. Reducing constraints improves flow and increases the throughput of the system. The process with the highest cycle time after improvement is 6 secs. This translates to throughput improvement from 150 units per hour (24 secs) to 600 units per hour (6 secs). This translates to a 400% throughput increase.

This is the amazing power for reducing the *mura* and constraints and is the initiative the organisation must constantly be focused on to achieve excellence in their manufacturing operations. The activity of reducing the mura in the production system must be carried out as a continuous improvement journey to achieve perfection in the flow and throughput results.

After reducing the constraints, as shown in the figure below, the production line can finally operate at 6 secs cycle time, which is equal to 600 units per hour. This should translate to an output of 6,000 units per shift (10-hour shift) and 12,000 units per day (two shifts per day). However, the problem was that the production supervisor was not able to deliver the 6,000 units per shift, delivering below 3,800 units per shift, which is only 63% output of the potential 6,000 units at 100%.

Figure 8: Before and After Reducing the Mura Waste

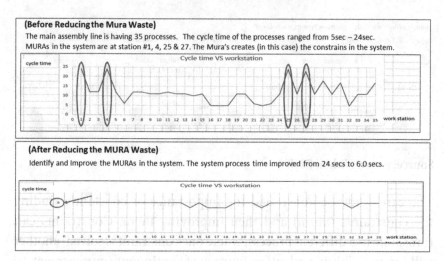

Source: Azlan Nithia, 2023

A team of engineers and a production supervisor did a 'constraint' study in the system, looking for the next *mura* or the next constraint. The team discovered the frequent production line stoppages due to a shortage of parts supply from another sub-assembly production line. The sub-assembly line is only delivering 380 units per hour to the main production line. Therefore, the main production line is capable of producing 600 units per hour, but because of the shortages of part supply from the sub-assembly production line, the overall system throughput was constrained to only 380 units per hour.

The team of engineers shifted the focus to improving the sub-assembly production line. Look at the figure 8, sub-assembly processes. There are 12 sub-assembly processes or stations. The stations 1, 3, 5, and 12 are not able to meet the target expected at 600 units per hour. The team was able to improve the output of those four stations by producing 600 units per hour. After the sub-assembly constraints were improved to achieve the throughput of 600 units per hour, it was now able to supply the required 600 units per hour to the main production line. Finally, the total system throughput was improved and able to produce 600 units per hour.

Figure 9: Mura Actual Case Study

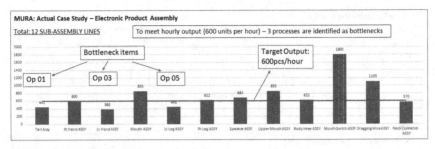

Source: Azlan Nithia, 2018

The case study demonstrated the importance of constantly finding the constraints (*mura*) and then improving them. Constraints can transmute to a different process, where one constraint is solved, another will emerge. This is a relentless journey of continuous improvement, to continuously identify and improve the system flow and improve the throughput.

Muri (Overburdened)

The third 'M' is the *muri*, which means overburdened. When a machine or person is overburdened, it can result in *mura* and *muda* wastes. *Muri* occurs when the workers or machines are pushed through their natural limits which can lead to additional workplace problems.

The performance of an overburdened machine or person cannot be sustained over time. The overburdened machines, operated beyond the natural limits will break down when the machine breaks down, creating shortages of parts and impacting the target output of that process. Thus it creates unevenness in the part flow to the next process and reduces the throughput of the system. Similarly overburdened people being worked beyond their natural limits will also break down. This translates into employee absenteeism. Therefore, it is important to ensure *muri* are not created in the production system as well as in the administrative processes.

All machines in the production must be operated per the machine's normal operating procedures. It is a good practice to implement the OEE (overall equipment effectiveness) for all machines used in production. It is important to maintain all machines and tools at optimum operating conditions so that unplanned breakdown is avoided. One method of achieving this is by establishing preventive maintenance procedures for all machines and tools used in the production processes. The production team and technicians must ensure a constant adherence to preventive maintenance (PM) procedures. Implementing autonomous maintenance is very effective in preventing unplanned breakdowns. A well-maintained machine will deliver a consistent performance.

In the human environment, Muri can be identified with the lagging employee absenteeism indicator in the twenty-first century, more and more people end up at home with burnout, which means they are literally overburdened up to a point where they cannot perform (Westendorp and Van Bodegom, 2015).

Apart from absenteeism, performance as a result of arousal levels is like a parabola in which performance increases when the arousal rises but only up to a certain point. After that, more arousal (now perceived as stress) leads to lower performance (Yerkes and Dodson, 1908).

Tools that help to reduce people-related *muri*:

1. Every process and operation must have a well-defined SOP (standard operating procedure) that visually explains the job and activity of the operator during the process.
2. The team board must visually display the 5S, standard work, hours required versus available to perform the task and absenteeism. This gives an indication of any overburdening in any production areas.
3. Production supervisors must constantly be engaged with the workers to receive feedback and listen to the problems being faced by the workers.

4. Managers must periodically have small group discussions with their employees to receive feedback and ideas to improve their work areas.

5. Not engaging the employees is a waste, the waste of unused employee talent. Employees must be actively involved in all the improvement programs and encourage them to participate in daily improvement activities in their work areas.

'*The Andon system* (jidoka*) reduces* mura *and* muri.'

This system to stop and fix the problems is one of the core pillars of TPS and is called *jidoka*. This action stops more defects from being made, prevents the defective parts from flowing to the next process, and triggers the problem to be resolved immediately by the line technicians.

Andon System

The Andon is the concept of Jidoka, which is one of the core pillars of TPS, to stop and fix the problems. At Toyota, the Andon concept starts with a cord hanging above the production line. Whenever an employee encounters a problem, the cord is pulled and the production line is stopped across the entire production line. All employees in the line know that a problem has occurred, and a visual or alarm sound system is triggered, telling them where the problem occurred. With the line being stopped, the employees can focus on immediately solving the problem. Stopping the line also ensures that defects are not created and passed on to the next stations in the production line as the defect is contained at the station where it was created. This prevents the multiple products or batches from being infected by the defects.

It is very useful to implement the visual Andon and alert systems in every process so that attention can be triggered by the operators who are running those processes in cases of interruptions to their outputs. These Andon systems must be supported by a good technician with prerequisite support and skills, who can take immediate action to keep the process or machine running effectively and must know the solutions to minimise downtime. The usual failure in Andon implementation is the help or support system. When the operator triggers the Andon, the help is supposed to arrive within seconds or a couple of minutes, but sometimes help does not arrive or help arrive very late. This poor response will demotivate the operators, and they will stop using the Andon, resulting in the failure of quick problem-solving or quick actions. What happens when the operator stops using the Andon? The output gets impacted because the machine stopped and not producing parts or the machine could be still running but is producing bad parts. In either case, it is bad for the company because it either creates a shortage of parts, or defects are discovered in the later processes or the final process. By then, it is already too late as throughput constraints are already created in the system.

The Andon system must be taken very seriously by everyone on the shop floor, including the area manager. Any failure to respond quickly to an Andon light must be treated as a major system failure on the shop floor and be treated very seriously by the organisation's leadership team. The technicians must be well trained, have experience, and have the necessary tools to take quick action to processes he or she is being assigned. It is useless to implement the Andon system but not able to respond effectively to solve the problem. Effective Andon response systems are normally lacking in many companies.

Value Stream Mapping (VSM)

The VSM is the process of carefully studying the activities carried out in the organisation with the objective of creating value and eliminating *muda*. It is an important tool for the leadership team to reimagine and reconfigure manufacturing operations to create value across the organisation and the supply chain. The VSM also realigns the organisation's workflow (creating value) which is key for any organisational intervention and transformation to achieve sustainable competitive business advantage. The VSM activity is generally done every year, and it is a continuous journey to reduce inventories and eliminate non-value-added activities. The VSM activity is a key driver of many improvements in the organisation—inventories, lead time, throughput, and reduction of overall product cost.

This VSM is generally done in a kaizen activity carried out by cross-functional team members. In this VSM kaizen activity, the team will observe and identify all the value-adding and non-value-adding activities currently done from material delivery through production processes, completing the product and finally delivering to customers. It is done by direct observation and the recording of every single activity, the material flowing in the system, and production processing, including inventories and works in progress (WIP). It is basically looking at the whole picture from ordering raw materials and finally delivering the product to the customer. VSM is a lean tool that is used to identify all the value-adding and non-value-adding activities and materials by doing a value map of the process from the supplier (receiving raw material) to the customer (shipping the finished product to customer), highlighting the flow of material, product, information, inventories, waiting, defects, reworks, delays, and all the non-value-adding processes being done in the system. This mapping process is commonly called VSM kaizen or flow kaizen.

In today's businesses, the application of value stream mapping, also referred to as 'visualising the flow' or 'mapping the flow' process, is not limited to the assembly line in production. With proper implementation, value stream mapping fosters a culture of continuous improvement that has been proven effective in manufacturing, information technology, engineering, financial, human resources, legal, and marketing services. Much of lean thinking in knowledge work starts with applying value stream mapping to any work where there are repeatable processes. It must always be done from end to end, not for a single process; it is about looking at the whole business. Although the technical value stream mapping definition varies by industry, its primary concepts have moved beyond manufacturing to be an effective tool for improving processes across all business functions or processes.

VSM is about discovering where we are today with a current value stream map (or value creation) and using the team or teams to create an ideal future state value stream map as a target for creating future state maps to work towards the journey of continuously reducing non-value-added activities in the system to achieve the ideal future state of increased value creation.

The VSM activity (or VSM kaizen) documents every activity, material being used, and materials in the inventories in every process and equipment used throughout the system. The data (activities and materials) is further analysed and classified as value-adding (for the customer) or non-value-adding for the customer. All materials held in between the process and in the inventories are classified as non-value-added (NVA). However, customers do not care about the inventories.

The VSM is an important process that every organisation must carry out to establish the amount of NVA that exists in the system and then establish opportunities to reduce those NVA. This is one of the most powerful and easy-to-use mapping tools and can lead the organisation to a rapid and significant improvement in business performance when the teams implement all actions following the VSM mapping exercise. The total VA and NVA percentage is recorded and established as the baseline for the next VSM kaizen. This VSM process is repeated every six to twelve months. Once all the VSM actions are identified to reduce the NVAs, the next step is to implement all the actions (verifiable actions) and ensure it is executed successfully and the gains are captured by constantly auditing the new SOP (operating procedure) to ensure new methods are followed.

The organisation leaders must actively and diligently be involved in helping the VSM teams to complete the actions in a timely manner. The leadership must take a personal interest in understanding the NVA and ensure the VSM teams are actively implementing the actions, sustaining the results and doing a weekly or monthly update of the status to the leadership team. Any organisation claiming to be on a lean journey but does not actively have a VSM exercise nor focus on reducing NVA is missing the core understanding of lean and TPS strategies.

The figure below shows an example of a nine-step process flow operation in a product assembly factory. The lead time taken from the time the material preparation was received to the time the products were ready to be shipped out of the factory is 360 minutes versus the actual value-added process time of 39 minutes.

Figure 10: Value Stream Mapping (VSM)

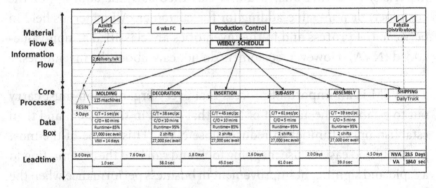

Source: Azlan Nithia, 2018

The value-added percentage or the index can be calculated as below:

Total Lead Time (LT) = 360.7 minutes. Total Value-Added (VA) Time = 39.3 minutes

The VA index (%) = 360.7 / 39.3 x 100 = 10.8 %

It will be shocking for any organisation to discover after the VSM kaizen is completed that 90% of the organisation's activities are not adding value to the product or to the consumer. Most of the companies will be equally surprised to discover that a VSM kaizen exercise would reveal their value added is less than 5%. These are common VSM discoveries. However, this result must be positively viewed as a big opportunity to improve current performance, especially in inventories, rework, lead time, throughput, quality, and cost.

The VA % or the VA index is a good indication of the opportunity that exists in the system currently, and the organisation must seriously focus on reducing the non-value-added (NVA) activities in the whole system. Most often, organisations make a big mistake by improving the VA portion of the processes, where the focus, instead should be to reduce the NVA activities (this is the *muda* waste). The VA is what the customer is willing to pay for. For example, when transporting parts from machine A to machine B, it is obvious that there is no change in the product form. This is classified as NVA. By keeping the parts in the inventory, there is no change to the product, again classified as NVA. Similarly reworking a defective part is NVA as well. Customers are not willing to pay for these activities as they do not see any increased value in the product.

The parts that are held in the inventory, storage, or work in progress are considered NVA because it is a waste to keep parts in the storage and inventory as it delays the flow of parts and results in a longer lead time. The first step to reduce NVA is to focus on reducing the inventory quantities (reduce them or at most a preference of elimination where and when possible) in every process. If a buffer inventory is required to manage certain constraints in the system, then there must be a very good reason. An example would be to reduce the bottleneck in the next process and improve flow. Any kind of storage of parts or material in the inventory must be done with a Kanban visual control system and with the minimum-maximum quantity inventory control which is established strictly to meet on time the customer deliveries in full. The organisation must be highly concerned with any kind of inventory being held in the manufacturing system. This is extremely important to ensure the production is operating with the smallest lot size or batch size with a lot size of one being the ultimate goal. The lean concept encourages manufacturing operations to continuously strive towards achieving the 'one piece flow' as a relentless improvement journey to achieve excellence in manufacturing. It is a continuous endeavour for the reduction of the inventory quantities in the production processes, improving the flow, and speed of flow which results in smaller lot sizes.

The VSM process should be repeated every six months (or at least once a year). All improvement activities must be derived and driven from the VSM outcome. After the VSM kaizen is completed, clear improvement objectives must be established. This is how more kaizen events are organised to realise the results expected from the VSM kaizen. All related actions must be tracked to ensure implementation and accomplish results as committed during the VSM kaizen exercise. The organisation should assign teams to implement all the improvement actions. This must be done in well-organised kaizen workshops focusing on ideas for improvement. These actions must be implemented using the kaizen approach and all the actions tracked through an activity called the 'make daily improvements' (MDI) activity.

CHAPTER 6

The Journey of Continuous Improvement

Early Detection and Immediate Action

The concept of 3 NDs (no defects):

1. *Make no defects (make good parts)*
2. *Send no defects (send good parts)*
3. *Receive no defects (receive good parts)*

These 3 NDs must be developed into an operational culture and instilled in everyone on the shop floor and the employees must be empowered to practice this without any fear. This culture of 3 NDs cultivates the system of early detection and immediate action which will result in preventing defects from being made; this prevents *mura, muri,* and *muda.* In any manufacturing industry, the concept of early detection, taking immediate action, and preventing defects from happening again is an important quality mindset requisite in a high-performing organisation. The A3 problem-solving process must be part of the organization's DNA.

'Early detection + Immediate action = Defect prevention'

The Pride of Workmanship

There must be a clear flow of problem-solving. Operators need to have a complete understanding of normal conditions. So whenever there is a gap, they know there is a problem. The operators or the technicians must learn to see the parts and the equipment they are using and be able to recognise when they have a problem. The first indication is when a part is defective, revealing that something is going wrong in the process. The operators and technicians must know if the conditions are normal or not. To appreciate this, they must be knowledgeable of what is an abnormal condition and be empowered to immediately stop the machine from making any defective parts. This is the core requirement of implementing pride in workmanship; it empowers the operators and technicians to take immediate action and to stop the production line or the machine whenever a defective part or an abnormal condition is found.

The '3 NDs' is an important work culture that cultivates the pride of workmanship in every employee. Employees must have pride in what they do and what is done is ready for the next process. This pride of workmanship culture applies to both manufacturing and administrative processes, in other words, the whole organisation. The employee who is doing the work has the responsibility for not making a defect. If the process or the method is not capable of making a good part, the employee must quickly raise the Andon for help and stop making defective parts—*do not make defects.*

In case when defective parts are inadvertently made, it is the responsibility of the employee to identify those defective parts and label them as defective. Those defective parts must not flow into the next process until they are reworked and confirmed as good parts at the identified station where the defect was created. The containment of defective parts at the source where it is created is extremely important—*do not send defects.* If a defective part is moved to the subsequent process, it only creates non-value-added work on a defective part. Ultimately these defective parts would either require a lot more reworking in the next process or be rejected (beyond the ability to be reworked). The employees must be trained or the process inputs must be designed (mistake-proofed) so that it is possible to detect any defective parts coming into the process—do not receive defects.

When a manufacturing or production group is highly focused only on output, measured by the number of units made by a process and highly driven by the quantity to deliver every shift, this will coerce the work behaviour of 'quantity first and quality second'. This behaviour will exist on the shop floor as an unspoken way of doing things in production areas. it is very prejudicial for the organisation because it gives the impression that the company is only interested in quantity. In this kind of work environment, the workers and supervisors are recognised for the quantity they deliver. The defective parts that were sent to the next process will continue till the final process is completed. Here the quality inspections discover the defect and reject the part at the end process.

This is also the reason why the local efficiency measurements measuring individual process efficiency are not effective. This inadvertently drives poor employee behaviour at the workplace and negatively impacts the teamwork driven by common goals.

So who is responsible for those defective parts? Who made them? Could it be known where and when the defect was created? How many defects were created and how many products—lots of produced parts—are now contaminated with this kind of defect? These are common problematic questions arising in many companies which personify the lack of 'pride in workmanship'.

To instil the 3 NDs mindset is not an easy task as it requires a new and different workplace culture. Start with one ND at a time; start with the first—do not make defects. This will require an effective Andon light or siren system, quick response from the technicians to solve the problems quickly and a strong commitment from the front-line leadership. If this first ND—do not make defects is not effectively implemented, then the other two NDs will fail. Immediate action must be taken as soon as a defect is detected by the operator. The first step is to determine the process or the machine that makes the defects. Subsequently trigger the Andon for help.

The 7-Step sequence of events when a defective part is produced and detected:

1. A defective part is detected by the operator.
2. The operator immediately stops the machine or the process that is making the defect.
3. The operator triggers the Andon for help.
4. The operator isolates those defective parts produced from the good parts.
5. The technician arrives quickly (within seconds or minutes), solves the problem, and confirms the process and machine are capable of producing good parts and hands over the machine to the operator.
6. The operator restarts the machine to produce non-defective parts.
7. The front-line leader must decide what action needs to be taken to rectify those defective parts that were produced.

The above is an example of a good manufacturing practice that is required on the shop floor to operationalise the behaviour of *early detection and immediate action*. This is one of the important 'people behaviour' and 'engagement' criteria that must be cultivated to achieve manufacturing excellence. For a non-defective condition, establish what needs to be inspected.

It is extremely critical to ensure that the engineers responsible for the machines and the processes are deeply engaged in the daily production performance monitoring for any new machine or process being released into production. There must be a system or procedure in place that holds the engineers accountable for 'clean releases' of machines and processes to production and complete the daily tracking till the machine and process achieve consistent quality and output performances. I have come across numerous companies, whereby the engineers would release the machine and process to the production operator with its inherent performance still questionable. This imposes a heavy burden on the operator and the production staff to meet the expected quality and hourly outputs. For that reason, it is important to have a procedure that requires production personnel involvement in every machine and process design, as well as a buy-off and handover procedure between the engineers and the production personnel.

The Concept of Jidoka

The term *jidoka* (a Japanese word) is used in the TPS (Toyota Production System) which can be defined as the 'automation with a human touch'. The word *jidoka* traces its roots to the invention of the automatic loom by Sakichi Toyoda, the founder of the Toyota Group. The automatic loom is a machine that spins thread for cloth and weaves textiles automatically.

Before automated devices were commonplace, back-strap looms, ground looms, and high-warp looms were used to manually weave cloth. In 1896, Sakichi Toyoda invented Japan's first self-powered loom called the 'Toyoda power loom.' Subsequently he incorporated numerous revolutionary inventions into his looms, including the weft-breakage automatic stopping device (which automatically stopped the loom when a thread breakage was detected), the warp supply device and the automatic shuttle changer. Then in 1924, Sakichi invented the world's first automatic loom called the 'Type-G Toyoda automatic loom (with non-stop shuttle-change motion)' which could change shuttles without stopping operation (this is the concept of zero loss time during changeovers).

The Toyota term '*jido*' is applied to a machine with a built-in device for making judgements, whereas the regular Japanese term '*jido*' (automation) is simply applied to a machine that moves on its own. Jidoka refers to 'automation with a human touch,' as opposed to a machine that simply moves under the monitoring and supervision of an operator.

Since the loom stopped when a problem arose, no defective products were produced. This meant that a single operator could be put in charge of numerous looms, resulting in a tremendous improvement in productivity.

The concept of Jidoka is one of the core pillars of TPS, which is to stop and fix problems. At Toyota, this translates into the Andon, a cord hanging above the production line. Whenever an employee encounters a problem, the cord is pulled; the production line is stopped across the entire production line. All employees in the line know that a problem has occurred, and a visual or alarm sound system is triggered telling them where the problem occurred. With the line being stopped, the employees can focus and immediately solve the problem. Stopping the line also ensures that defects are not passed on to the next stations in the production line. The defect is thus contained at the station where the defect was created. This prevents multiple products or batches from being made defective, and immediate action can be taken to rectify the problem—early detection, immediate action.

Jidoka's 4-Step Approach

    i.   Detection of the deviation from the specification or the SOP
    ii. Immediate stoppage of the production line
    iii. Fast action to fix the problem and start the line
    iv. Analysis of the root cause of the problem and implementation of preventive action so that the same problem will not reoccur.

Since the equipment stops when a problem arises, the operator can visually monitor and efficiently control many machines. As an important tool for this 'visual control' or 'problem visualisation', it is important to use a visual problem display board system and locate on the shop floor, 'visual and on board' that allows operators to identify problems in the production line (machines and processes) with only a glance at this visual Andon board.

It is important that the operational system is adequately developed and can execute early detection for immediate action all the time. The employees and the technicians must be well trained, equipped, and capable of practising this *jidoka* system of early detection and immediate action repeatedly. The employee culture of early detection and immediate action can happen only if the employees are good at problem-solving or have the ability to take immediate corrective action to keep the process or the machine running, manufacturing good quality parts with very little downtime or loss of time.

Make Daily Improvements (MDI)

Very often, companies claim kaizen outcomes or results cannot be sustained, resulting in the failure of the lean journey, where many reasons for failure are given by these companies. These failures to sustain the gains can always be traced back to the failure of execution. The MDI activity is an extremely important exercise to ensure all the kaizen actions are implemented effectively and efficiently. It involves the responsible process owners. The kaizen team must audit the area and processes daily with the process owners to ensure all actions are implemented as planned and get the employee feedback for any adjustments to the implementation plan. This is the meaning of making daily improvements.

The kaizen improvements fail and cannot be sustained because of the failure of not doing any MDI activity. Companies must cultivate a 'gemba' culture, a culture whereby the management team is willing and happy to walk the shop floor daily. The product value is created in the production processes (on the shop floor) and not in the offices.

The management team must take a serious interest in supporting the teams and ensuring all kaizen action items are effectively implemented and deliver the results. A weekly review, a thirty-minute update by the kaizen team to the management team must be made compulsory for both the kaizen team and the management team. These weekly kaizen updates must start and finish on time to indicate commitment and discipline. Starting the meeting late and finishing it late indicates that these update meetings are not important and not well organised.

Daily teams must walk the shop floor and observe any new work method, looking for deviations or difficulties faced by the employees involved in the new work method(s). Encourage the employees to record the difficulties they faced, corrective actions taken and improvement ideas mooted. It is recommended to provide a visual board or flip charts in the work area so that the employees can write down and be visual. This is called the MDI kaizen newspaper. Look for ways to help the employees solve these problems immediately.

For example, the process engineer must walk the shop floor to review the performance of the processes that he or she is in charge of constantly looking for process improvements. Equally important is safety (including the environment and health conditions in the area). Irrespective of how small problems can be, look for improvements.

The supervisors, managers, and the organisation's general manager must set an example by walking the shop floor daily and asking the employees about the problems of the day and how they were solved. As for the human resources manager, where are the people that they are serving? As we know, most production workers are on the shop floor. How often does the human resources team walk the shop floor? A new organisation culture must be created to operationalise a new set of people's behaviour.

Sustaining versus Continuous Improvement

The biggest and most difficult task in any transformation or improvement program or after a kaizen event is to sustain the gains. Most of the lean and kaizen improvements evaporate quickly; some even are lost within a week or two after the improvement kaizen event.

The discussion on sustaining improvements and gains is a hot topic amongst the leaders in many organisations embarking on the lean journey. Leaders very often complain that they are not able to retain the improvements set up. Consequently there is the need to re-kaizen or the need to regroup the team to redo the activities over again and again.

Question:

- *Why is sustaining the improvement important?*
- *Why not continuously focus on improving instead of sustaining?*

Figure 11: Sustainable Continuous Improvement—the Step Journey

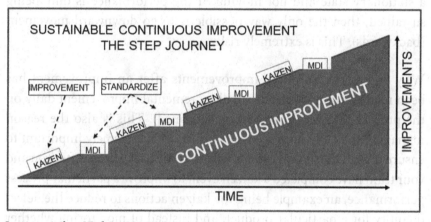

Source: Azlan Nithia, 2023

I have personally seen numerous organisations struggling to sustain the improvements every year. The performance slides backwards down to the old performance levels with the improved results unsustainable. The leaders have lots of reasons and excuses. Often they blame poor follow-up and lack of commitment among their subordinates.

If you and your team truly believe in continuous improvement as an organisational work culture, then why sustain a performance standard? Instead of sustaining, why not drive continuous improvement—small incremental improvements (micro-improvements)? The danger of sustaining is sliding backwards, but the power of small incremental improvements means moving forward. For that reason, sustaining the improvements becomes a spin-off of the concept of continuously improving.

> *Sustaining = Not moving up or down (not improving)*
> *Continuous improvement = Always improving*
> *The weapon for manufacturing excellence*
> *is continuous improvement.*

Think of small improvements instead of sustaining the improvements (made earlier) that are already in place. Sustaining means you are in a stationary state and not moving. If the performance is only being sustained, then the only way possible is to go downward movement (backwards). This is extremely risky.

Why operationalise micro-improvements after an improvement has been made? It simply means small incremental improvement daily on the process that was improved or kaizen(ed). This is also the reason managing the daily improvements (or MDI) is extremely important to ensure that all the improvements will not slide backwards. After you and your team have completed a kaizen event to improve a particular process performance, an example being the kaizen actions to reduce the defect quantity for a particular product, and instead of monitoring whether the process team is sustaining the performance, do the daily *gemba* observation and look for improving it further. Those who are in charge or responsible must do the daily *gemba* observation together with the Kaizen team members. This must carry on for at least thirty (30) days. During this period, clear SOPs must be developed, all the requirements must be completed, and consistent performance and results must be achieved daily and every hour.

Continuous improvement is the most powerful manufacturing excellence 'weapon'. This must be the *bedrock* in any organisation's DNA and a daily work culture strength across the whole organisation. Everyone walks the talk of continuous improvement in all the work they do by demonstrating that every day is a better day.

Manufacturing Excellence—the 3M House

The struggle that many organisations face is not in where to apply lean or agile tools but in effecting the organisational and behavioural change required in a successful transformation. The new change in organisational performance will not happen if the people's culture remains the same because it is the people that impact the change. The 'manufacturing excellence 3M house' explains the core foundation (people culture) that is required to continuously reduce the 3M waste in manufacturing to achieve manufacturing excellence. The organization's people culture and its capability to impact change are the core foundations for achieving manufacturing excellence.

However, while embarking on a transformation journey using Lean methods, many organizations fail to realise that without a change in organisational culture, lean methods or manufacturing excellence are unlikely to be successfully adopted to augment the business value of the organization. Lean methods or lean manufacturing is not a program or a set of tools, but the transforming of an organization's culture. It is about leadership commitment; it is a relentless journey without a finishing line and a pursuit of continuous improvement. Often organisations fail to articulate a compelling vision for this change, one that can be consistently communicated to the employees and readily internalised by everyone at all levels.

What is manufacturing excellence? What do the manufacturing excellence experts say about the organisations that are serious about achieving manufacturing excellence?

- *Larry E East (2016) in his book* The 12 Principles of Manufacturing Excellence *states, 'It is not the flavor of the month; it is continuous improvement (CI)'. It is forever. It must be the operating strategy for the business.*
- *Andrew Miller (2014) in his book* Redefining Operational Excellence: New Strategies for Maximum Performance and Profits Across the Organization *states, 'Operational excellence is the relentless pursuit of doing things better. It is not a destination*

or a methodology but a mind-set that needs to exist across an organization. It is about empowering employees to use judgement on the front lines'.

Numerous cost reduction programs are usually launched to achieve transformational initiatives that would apply lean methodology to reduce costs, process complexity, and eliminate waste from the system. The complexities involved in managing lean transformation programs have made the success rate or the sustaining of such transformations to be poor, especially the ability to sustain the lean gains. The key reason for the low success rate is that many organisations deploying lean manufacturing use them as a tool to achieve cost reductions as their only prime objective. The decisions made by the leaders very often overlook the organisational, behavioural, and cultural aspects of this lean transformation. The leadership team must understand that people culture and engagement of employees play a very critical role in the success of lean adoption as an organisational culture and set an organisational ambience for other transformational programs to succeed.

'It is all about the leadership team's commitment, the people's engagement to drive improvements, and create them as an organisation work culture at all levels.'

People Culture Is the Foundation

The future of manufacturing is rapidly changing, morphing into a digitally powered automated manufacturing environment that will require highly skilled talents while becoming less labour-intensive. Therefore, the skills needed are changing, requiring the ability to manage complex tasks, and the aptitude to constantly learn and apply new knowledge. The kaizen philosophy requires step-by-step improvements, process refinements, and continuous enhancements, which are the surest and fastest route to achieving optimum productivity gains and quality improvements. Making small step improvements, done in many processes, will lead to accumulated significant improvements in productivity, quality, lead time, and cost.

The change and improvements have to be an ongoing process. It is good to establish a good change process communication system in the organisation. The only thing that will not stop changing is the change itself. The best change is always when people think they did it themselves. High involvement is good but never becomes cumbersome, and it should not interfere with the people's success in their regular function or role.

People usually will not mind new changes made in the workplace if they are used to the idea that the management will always give them an opportunity to be involved, and they too have an impact on the direction of the change. It can even be as minor as asking the people for their opinions of the new direction or the new change. This will improve the participation level and involvement of the people to execute the new direction from its early implementation stage to full implementation.

Figure 12: Manufacturing Excellence—the 3M House

It is all about talking to the people, communicating to them openly (having a structure in place on how this communication will be done at all levels in the organisation), and letting them feel that they are part of the new change process. It is important to recognise that under every successful new change implementation, there is lots of hard work, and it has a lot to do with good and effective communication.

Develop the measurement systems and build them into the change process that visually tells the performance to the people whether they are making progress or not, based on the new or revised targets. There must be full alignment between all parties which includes the negative consequences of not following the new SOP and rewarding those who positively deliver good results.

Creating a work environment, in which employees feel as if they have the power to initiate change, is also positive and a tribute to the work culture. However, employees find themselves caught up in changes that others are initiating.

Innovation and automation-minded managers are generally quick to resort to buying new automated machines and robots to create islands of improvements, but it may not create total system value or actual enhancement in end-to-end productivity gains. It may in fact impede changeovers and reduce flexibility. Leaning out the processes through a VSM kaizen should always be the priority. Remove the 3M waste first and the implementation of automation or robotics should be last, if not, the engineers would have only automated lots of non-value-added waste.

The key to achieving manufacturing excellence in a competitive business environment is having the right people and organisational culture in place that is always ready to learn and apply, able to deal with complex challenges, and constantly improving. To adapt and manage change effectively, continually realigning and reassessing the organisation's people culture is crucial. It is all about the people and the organisation's agility that is required to drive continuous improvements and create a change for the better.

The changing demands of the kind of workforce, talent, and skills needed, as we match into the new talent requirements of the future, is a learning organisation that can adapt quickly to changing requirements and roles.

The organisational needs to seriously consider people's talents

1. Employees should have good presentation skills and the ability to articulate new ideas effectively and quickly
2. Creativity and the ability to solve problems effectively and generate creative solutions are needed.
3. Employees must have the ability to critically analyse and think through the various options to develop better ways of doing work
4. As organisations embrace and implement the requirements of smart factories, the employees must be able to autonomously work with digitally connected systems, machines, and digital feedback outputs to make decisions.
5. Teamwork is critical. They must be able to work across various functions to efficiently complete the task. They should be agile to work on multiple processes and adaptable to changing customer requirements.
6. Take responsibility, accountability, and constant improvements.
7. A human resources (HR) team should understand the importance of the above and effectively provide the support to make it happen.

The 'manufacturing house—3M waste' consists of two very important people's foundations:

- everyone continuously improving (and learning)—every day a better day
- daily *gemba* kaizen, *jidoka*, and problem-solving

Then those two core people foundations are the three key pillars of 3M:

- *Muda* waste (to discover, reduce, and eliminate)
- *Mura* unevenness (to discover, reduce, and eliminate)
- *Muri* overburden (to discover, reduce, and eliminate)

On top of these three 3M pillars (*muda*, *mura*, and *muri*) sits the activities—'constantly identify, reduce, and eliminate all the 3M waste in all the processes and activities', finally it is the outcome that leads the organisation to achieve the 'manufacturing excellence'.

The Four Key People Culture Foundations

There are four key foundations that must be developed and operationalised as part of the organisational people culture (like the company DNA). These four components form the core requirements for any high-performing organisation to be successful in implementing smart manufacturing and the factories of the future. This foundational culture will ensure sustainable business growth and profitability and lead the industry.

1. Continuous improvement

Employees in the organisation must practice the work culture of continuously improving the processes and be engaged in daily improvement activities. It is important for the organisation to embrace, cultivate the learning culture, and ensure that every employee is engaged in finding new knowledge and implementing new ideas to enhance performance.

The concept of 'every day a better day' is the key concept of continuous learning and continuous improvement. This is a work culture that must become the core foundation for all the employees and the management team in the organisation. If this is not developed or operationalised, then all the other improvement programs will not stick and cannot sustain the gains or improvements. The improvements will be lost very quickly after any kaizen event is completed.

2. Problem-solving culture

Discovering daily problems, highlighting them, and taking immediate corrective action is at the core of the problem-solving culture.

Problem-solving must be practised at all levels. The ability to effectively and quickly solve problems is a key requirement of high-performing teams and employees in an organisation. If the organisation does not have teams and employees who can effectively and quickly solve daily and chronic problems, it will not be possible to achieve high performance. Quality improvement and effectively solving problems are the key requirements to be successful in any industry. The quality improvement process is an endless journey, continuously pursuing the goal of zero defects or ultimate perfection.

Therefore, it is essential that the employees engage in daily activities and constantly improve quality by identifying problems and using a simple A3 methodology to systematically solve the problem and prevent the problems from reoccurring again. The goal should be to solve the day's problem within the same day. It is important to visually track the percentage of the problems that were solved within the same day (strive towards 100%) and review it the next day during the daily *gemba* problem-solving activity (this should be conducted early in the work shift). These daily problems must be tracked and recorded in the kaizen newspaper, a flip chart that is used to write down the problems and to show the immediate actions that were taken (located in the process). These problems must also be tracked to ensure they do not recur. For those recurrent problems, an A3 problem-solving kaizen must be conducted to further break down the problem, find the root cause, and implement the actions to prevent it from happening or occurring again. This is the team's relentless focus to ensure the same glitches will not be repeated.

The A3-DD Methodology for Problem-Solving

Problem-solving must be made simple so that everyone in the organisation can solve problems quickly and effectively. There are numerous problem-solving tools available in the market. Normally it requires very lengthy and completed processes to find the root causes or to solve the problem. As a result, very often companies will assign one or two highly trained individuals who are specially trained to use these complex problem-solving tools or methodologies. This kind of complex approach to solving a problem should not be encouraged. It is also very common for engineers to immediately jump into the causes; this is wrong and will result in difficulty in solving the problem permanently, especially those engineers who have been working in the area may infer that they know and may also use past knowledge to solve current problems. This is a dangerous behaviour towards problem-solving. Every problem must be taken very seriously, and one must always establish the root cause very precisely. Preventive actions must be completed to ensure the same problem will not reoccur.

The problem must be defined correctly. How one defines the problem will determine how it will be solved. There are big problems with many variables, and there are also small production problems which need to be solved quickly. The big problems (generally with several variables) may require more time and effort to solve. The big problems should be converted into a problem-solving kaizen event and turned into a regular kaizen project.

The simplest and most effective problem-solving methodology that I have personally used to successfully solve hundreds of problems is called the 'differential diagnosis'. This is a globally proven and powerful approach towards solving any kind of quality problem. This methodology is also known as DD in short. The founder of this DD methodology for solving quality problems was invented by the person known as Prof Dr Shrinivas Gondhalekar (aka Dr G). Dr G introduced his DD methodology in the book titled *Chronicles of a Quality Detective—developing differential diagnosis: a powerful approach towards solving quality problems* (Gondhalekar, 2005). The DD approach has also been successfully used in solving various kinds of administrative-related problems.

I have personally used this DD methodology for over fifteen years to solve various problems, and I have found this method to be the simplest approach that can be easily adopted companywide by everyone. This method also facilitates quick problem-solving.

Most often, engineers and problem solvers focus on why defects occur and therefore spend a great deal of time and resources analysing the defective parts or items. In most cases, the defect percentage will be very low, in the region of 5%–10%, even as the good parts or parts without any defects are more than 90%. This is how most of the problem-solving approaches work. Therefore, this has become a natural problem-solving behaviour used by engineers in most organisations, which is to look at the defective parts and find out why the defect occurred.

I have found it time-consuming to only look at defective parts to determine the root cause. With this in mind, organisations deploy very complex and sophisticated problem-solving methodologies. This requires certain individuals to be specially trained to become problem-solving experts in the company. I have found this approach, of having only one or two experts to solve a problem, very unproductive because problem-solving must be everyone's responsibility; not confined to certain experts only. To become a high-performing individual in a high-performing organization, the problem-solving skills are critical.

Figure 13: Focus on Why Parts Are Good

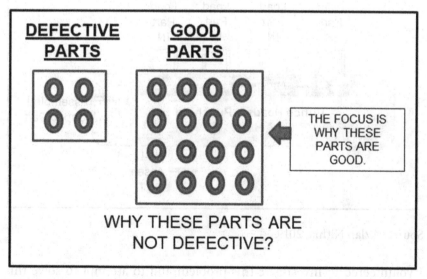

Source: Azlan Nithia, 2023

Problem-Solving Case Study

One of the processes in a company (let us name it company A) used for this case study called the 'load-punch-and-cap assembly' process, which produces a PVC part with a cap had lots of missing production orders. The production supervisor was complaining that less than 70% were good. This was after a lot of rework was done, and it required many rework operators. There are four machines, and each machine does its part loading, punching a hole, assembling the cap, and finally unloading the part. The general manager of this factory had approved the replacement of these four machines based on the equipment engineer's recommendation. The reason for replacement was justified as these four machines were over five years (old), produced too many defectives, increased the production cost, and lowered the productivity of this operation. The company was missing customer orders, and the situation had reached a serious level due to delivery shortages and customer grouses.

Figure 14: PVC Part Load, Punch, and Cap Assembly Machine

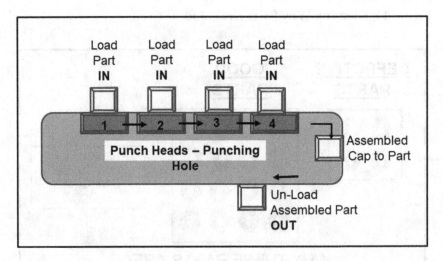

Source: Azlan Nithia, 2019.

I volunteered to investigate this problem and to attempt to solve this defective parts problem at all four machines. I requested a day to review the situation and give my recommendations. Meanwhile I requested that the general manager put on hold his decision to buy new machines.

The approach used is as follows:

1. The first and most important step is to define the problem precisely. An accurate problem definition is the key to successful problem-solving.
2. **Step 1—define the problem**. The defect was defined as the excess waste material on the part after punching (at the edges around the hole). This excess material was not acceptable and had to be removed manually after the punching and cap assembly process, which was done at another rework station.
3. The defect was the excess material at the edges around the hole for if there was no excess material around the hole, the part was considered acceptable. The problem definition is expressed as the <u>gap between the expected target results versus the current actual results</u>. The gap between these two (target versus current) is the problem. If this gap is closed, where the target equals actual, there will be no problem.
4. The defective parts could be reworked manually but require extra time and operators. The rework process is not easy as it also produces between 20%–30% defects (those that cannot be reworked).

Figure 15: The Gap Is the Problem

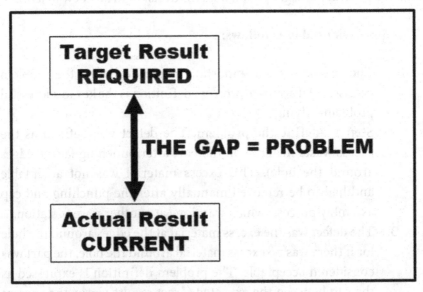

Source: Azlan Nithia, 2019

5. **Step 2—where is the defect present and absent?** The defect was only present around the hole area and was absent in all other areas. The defect was discovered at all four machines but only at machine 1. Punch heads 2 and 4,'s parts were found to be good, and punch heads 1 and 3 were found to have defective parts (excess material around the hole).

6. **Step 3—after which process the defect is present and absent.** The defect was only present after the punch head process as no other defects were created in any other process. Therefore, a conclusion was reached whereby the defects were found only after punching the hole during the punch head process.

7. I observed all four punching machines and looked at all the parts from each punching head. There were a total of four machines, and each machine had four punch heads, therefore, a total of sixteen punch heads, two making good parts and fourteen others making defective parts with excess material. I also discussed with the machine operator to get inputs from the operators regarding their observations, which machines were

making good or bad parts. It is important to have a discussion with those directly involved in the process. This allowed me to gather additional information that might assist in solving the problem.

8. Each machine had four punching heads, and each punching head punched one part at a time. All the punching heads were in one straight line. All punching at the same time, therefore, one machine can produce four parts per punching cycle. Each machine was operated by one operator. The four machines' performances are as below:

   – Machine 1. This machine had two heads (heads 2 and 4) making good parts and the other 2 heads (heads 1 and 3) making defective parts.

   – Machine 2. This machine had all four heads making defective parts.

   – Machine 3. This machine had all four heads making defective parts.

   – Machine 4. This machine had all four heads making defective parts.

9. **Step 4—compare the good and the defective**. I compared the machines and punch heads producing good parts with defective parts, i.e. with excess material around the hole. Machine 1 with punch heads 2 and 4 were making good parts. All other punch heads were making defective parts (the hole had excess material). The question that I needed to ask why machine 1's punch heads 2 and 4 were making good parts. The focus is now required only on the punch head process because step 3 indicated that the defect was present immediately after the punch head process.

10. I asked the operator managing machine 1 if anything was done by anyone to the heads 2 and 4. The operator showed me the maintenance record book and pointed to the work that was done by a maintenance technician to head 2 and later to head 4. After reviewing the maintenance record book, it was indicative that the technician had replaced the punch cutter on head 2 and later changed the punch cutter on head 4 as well.

11. The next step is to speak to the technician who changed the punch cutter on heads 2 and 4. I had a discussion with the technician and confirmed he changed the punch cutters in punch heads 2 and 4. The reason he changed was that the cutter was cracked and could not perform the punching process. This was important information because the technician had fixed new punch cutters to heads 2 and 4. This confirms the premise that after changing to new punch cutters, the defect related to excess material at the hole disappeared at punch heads 2 and 4. It is always important to establish the clue. In this particular case, it was the punch heads 2 and 4 with new punch cutters.

12. I spoke to technicians and the operators managing the four machines asking if they would allow me to replace the punch cutters on machine 1 and on punch heads 1 and 3. After agreeing, I requested that machine 1's punch cutters be stopped and removed on heads 1 and 3. Replaced them with new punch cutters. It was a fast twenty-minute job. After testing the function of the new punch cutters, the technician restarted machine 1, now with new punch cutters on heads 1 and 3. The results were amazing; heads 1 and 3 produced a clean hole without any excess material around the hole.

13. Next, the solution was repeated on the rest of the machines 2, 3, and 4. After about one hour, all the machine's punch cutters were replaced, and the process was made to run again. All the machines were re-inspected and reconfirmed by the quality inspector, the technician, and the operator. All of them confirmed that all parts from the four machines were acceptable and were good with no more excess material around the holes. This means the reworked process can be stopped, and the operators currently doing the rework can be deployed to other useful work. I can now confirm that the problem is solved, and I can also recreate the defect again because I now know the root cause of the problem. I continued to observe all four machines for another hour to ensure the operator, technician, and quality inspectors were satisfied and confirmed the problem was now solved.

14. **Step 5—preventive action to ensure the same problem will not occur again.** This is an important step to ensure the problem will not reoccur. This requires a new SOP (standard operating procedure) that will drive a new work procedure to prevent the same defect from resurfacing again. The new SOP will ensure that the punch cutters are inspected by the technician on every shift to confirm they are in good operating condition, together with the operator checking them hourly. The punch cutter quality specification was also determined and added to the SOP along with the weekly preventive maintenance schedule for the punch heads in all four machines.

15. Summary of the 5-step problem-solving approach

The below figure explains the five-step approach that solved the problem and will prevent the problem from occurring again. The general manager confirmed that the machines were operating well and producing good parts. The purchase of the new machines was cancelled.

Figure 16: The 5-Step Problem-Solving Approach

THE 5-STEP PROBLEM-SOLVING APPROACH

➢ Step 1 - Define the Problem

➢ Step 2 – Where is the Defect Present and Absent

➢ Step 3 – After which Process the Defect is Present and Absent

➢ Step 4 – Compare the Good and the Defect. Identify the Solution

➢ Step 5 – Preventive Action to Ensure the Same Problem will not Occur Again

Important: Follow the methodology strictly. Do not make any short-cuts to the solution.

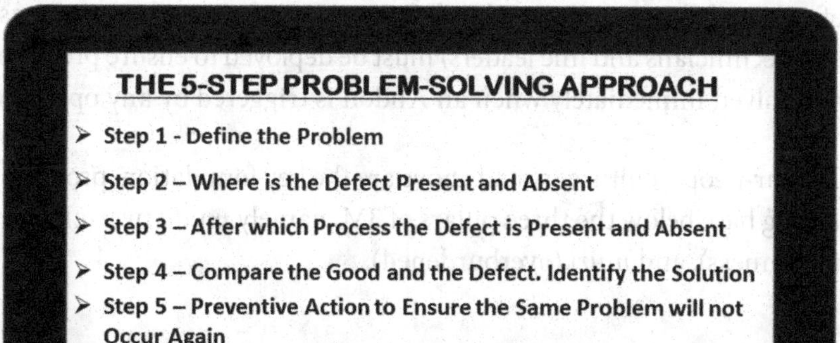

Source: Azlan Nithia, 2023

The Daily *Gemba* Kaizen

This is the 'small group-kaizen Activity' or SG-KA. Daily, the SG-KA team will walk and observe the work area to find opportunities to improve and take immediate action to implement those improvements. The SG-KA team members are selected from the line operators, line leaders, technicians, and quality personnel. This daily *gemba* kaizen is normally done during the beginning of the work shift for about twenty minutes, and then the team regroups to discuss the improvements, what actions to be taken, by whom and when. These improvements are then tracked and reviewed daily to ensure their successful implementation.

Generally, these small improvements are implemented daily. If the team finds a bigger improvement opportunity, which is beyond the team's ability to implement, the supervisor or the area manager will get involved in helping the team.

Constantly Improving the *Jidoka* (Visual Andon)

Stop and fix the problems. Do not make defects, do not send the defects, and do not receive the defects. A good shop floor response team (technicians and line leaders) must be deployed to ensure problems are resolved immediately when an Andon is triggered by any operator.

The four-people cultures stated above are the key foundation, providing a strong base below the three pillars of 3M, namely *muda* (waste), *mura* (unevenness), and *muri* (overburdened).

The Four Key People Engagement and Involvement Culture

1. *Continuous improvement—'every day a better day'*
2. *Daily problem-solving—simple and effective (A3)*
3. *Daily* gemba *kaizen (SG-KA)*
4. *Constantly improving the* jidoka—*visual Andon system and response time*

These four 'people culture' components must be developed, cultivated, nourished, and relentlessly reinforced in the overall organisational culture, embracing the way we do business and how we work. This must be applied across the whole organisation across all the functions in manufacturing and business processes.

Only by relentlessly pursuing and attaining all the key people concepts and having these concepts entrenched in the DNA of an organisation can the organisation achieve manufacturing excellence—which essentially is lead time and product cost reduction and throughput and productivity enhancement.

Visual Performance Management (VPM)

In every functional process and work area, it is important to create awareness of the area's performance using visual, digital dashboard displays (real-time performance displays) to the employees working in that process area and also for the management team to see performance status during their '*gemba* walk' (or the shop floor walk). The visual, digital charts should show current performance status versus targeted performance. This should be part of the connected factory or the smart manufacturing strategy. Real-time data allows the employees to take quick action to solve any issues that prevent the meeting of the targets.

It provides visual feedback to the employees on their ownership of the results (good or bad). These visual boards must also show any action taken to mitigate any problems that have arisen or unsolved issues requiring supervisor intervention. It also supplements the 5S process of visual workplace and visual flow.

VPM provides real-time information on process performance so that immediate actions can be taken. This enables continuous improvement and quick problem-solving, ensuring problems are not prolonged or carried over into the next shift. Quick response, to resolve any problems on the shop floor, is extremely critical.

The VPM must also ensure the illustration of visual material flow on the production shop floor. Every material flow must be clearly identified.

Some examples of material flow visual identifications (MFVI) required

1. raw material input
2. input (to a process)
3. accepted and rejected material
4. material that requires rework
5. material pending quality decision
6. output accepted and completed
7. All inventories (small or large) must have a *kanban* system (minimum and maximum inventory levels allowed). No inventory is allowed without a *kanban*.

The following are the daily and weekly key activities that are for VPM implementation to effectively drive improvements:

Operators—visually record hourly production data, production interruptions, material shortages, and equipment stoppage time. Utilise digital technology instead of manual.

Front-line leaders—end-of-shift meeting with operators at the production visual performance and digital dashboard to review output, and any issues and prioritise improvement activities for the next shift.

Managers—morning meeting with the leaders at the production visual performance board to ensure the availability of resources required for solving problems and those that require management help.

General manager—weekly meeting with department managers at the production visual performance board to review the week's performance, problems solved, preventive action taken, improvement ideas, and those requiring management help or support.

The supporting departments such as maintenance, quality, engineering, procurement, HR, and administration team members should join the managers and the general manager.

The management team should decide on the critical charts or measurements that must be displayed on the VPM digital board or the visual dashboard. This VPM board can be manually managed by the area employees or digitally managed by using large monitors. VPM must be kept very simple with very few words but have more visual performance trends and graphs to show whether there are improvements or otherwise and should be easily understood by the employees.

An example of the VPM digital dashboard (real-time data) is shown below.

Figure 17: Visual Performance Management (Digital Dashboard)

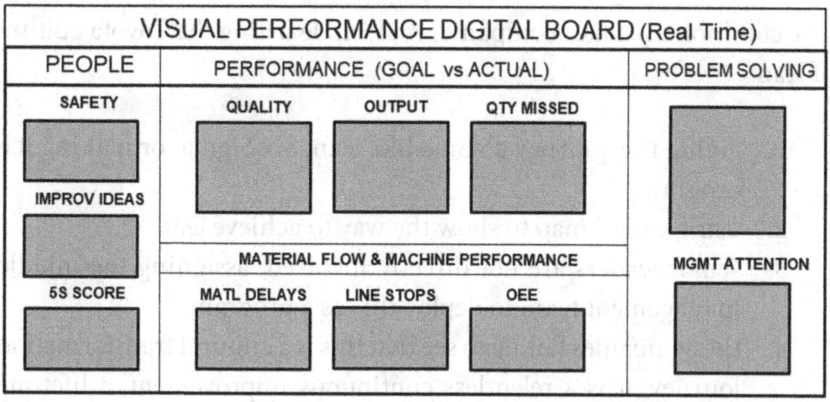

Source: Azlan Nithia, 2023

Implementation of the VPM boards requires management and leadership commitment to provide the required support, for example, a clear SOP with a support structure for technicians' response time to a machine down, an area leader's support system to resolve any material supply issues and supervisors' support to ensure problem-solving are completed quickly.

The leadership team's commitment must be visually seen by the employees as this drives an employee's commitment and passion to achieve better performance and generate ideas for continuous improvements. When a problem is not resolved quickly, the management team must ensure that the technical team is involved to resolve it and to ensure the same problems do not repeat again. Problem-solving and problem-prevention must be taken very seriously by everyone at all levels of the organisation.

The Toyota Production System (TPS)

TPS is a living system. It is not a toolkit or a road map but a culture of how we live it. You must live it to understand it as it is continuously evolving. I have seen many companies that find it overwhelmingly seductive to have a tool kit and a road map. Because of this desire, lean consulting companies feed on this need and happily provide their customers with whatever they want.

The common mistakes companies make in learning the Toyota culture of lean:

1. giving this journey a name like lean, Six Sigma, or making it a program.
2. using a road map to show the way to achieve lean.
3. senior leaders are not directly involved, assigning the middle management team to deploy this as a program.
4. the companies failing to see that this is a cultural transformation journey, it is a relentless continuous improvement, a lifetime with no finishing point.
5. senior leaders not taking responsibility for leading this cultural change, this is a difficult journey that requires.
6. a lean manager is assigned to deploy the lean journey (thinking it is a program) is not acceptable. This is not a one-person job, but it is an organisation's people transformation journey. Lean manager should assist in tool kit and capability training, namely *kanban*, 5S, problem-solving, SMED, and track the improvements of lead-time reduction, inventory reduction,

changeover time reduction, product cost reduction, throughput improvements, and visual factory (and visual performance) and organise the weekly updates with the leadership team.

This culture needs the top leader of the company to fundamentally build a new culture, engage in a deep understanding of the people involved in the processes and focus, first and foremost, on satisfying the customers. Progressing through the lean journey, companies mature from process-improvement toolkits to lean value-stream management; to employee engagement in daily problem-solving and self-aware leadership aligned to the right business problems. This means to journey through a business transformation that puts customers first together with developing people's culture and capabilities. Employees who are doing the work are constantly improving their work output by making problems visible, thus helping them to think about how to solve them. There are tools, whether they are *kanban*, 5S, or SOP, to establish a standard and to make any deviations from the standard visible to the employees in the workgroup. As a consequence, the work group develops the problem-solving skills needed to identify the root cause and solve the problem.

If these tools do not change the way people work and think about their own processes, the tools are a failure. If the leaders and managerial team do not understand how to use the tools to unleash the creativity and motivation of their people, the premise will arise, portraying them as not true leaders but just administrators in a bureaucratic system.

TPS is the company's pursuit of operational excellence which is also called the 4P model. This drives the intense focus and everyone's energy from senior management down to the operators on the shop floor to constantly find better ways to remove accumulated waste in the process and in their daily work.

The culture of respect for people and continuous improvement resides in an organisational culture. In retrospect, everyone in the organisation must have that external focus of adding value to the customers and shareholders.

Toyota's 4P Model

1. **Philosophy**
   This is the foundation to base the management decisions on a long-term philosophy even at the expense of short-term financial goals.

2. **Process**
   Constantly be focused on improving and eliminating waste, creating flow, and removing the 3M waste.

3. **People** (and Partners)
   Respect for people, constantly challenges them to better perform, grow their capabilities, induce creativity, and enhance the ability to always find better ways. Stimulate employees to have a creative spirit and encourage them to realise their goals or dreams (without losing the employees' drive) while constantly maintaining high energy levels.

4. **Problem-solving**
   Create a culture of continuously improving and learning. Train all employees to use the A3 methodology to solve problems daily and use preventive actions to ensure the problem does not recur. The people make all the difference in the organisation.

The three enemies of lean

There is a reason why the *muda* is there, and the reason often arises in conjunction with the two other enemies of lean, namely *muri* and *mura*.

The three enemies of lean are **interrelated** *and should be considered simultaneously in a never-ending quest to improve and achieve* **manufacturing excellence.**

Lean Agile Manufacturing and Culture

As shown in the previous sections, lean manufacturing is crucial to the deployment of integrated process efficiency in mass customisation. However, the introduction of lean manufacturing itself needs a reassessment of the organisational culture. This is because waste reduction often requires the emergence of a new work culture. Instead of pushing as much work as possible out to the shop floor, the emphasis should focus on how materials have to be brought to the line when production requires them.

The change to a lean manufacturing culture is profound. Attempts to introduce new cultures trigger powerful organisational defence routines that undermine the new culture. This ultimately stifles the transition of the lean process from the cell and shop floor stage to the value stream and value systems stage, necessary for mass customisation.

It has been argued that JIT is associated with basic techniques of inventory and production control, and TQM is a set of basic techniques that reduces process variance. HRM is a set of practices that shape the organisational environment in which the basic techniques are implemented.

A similar finding provided an indication of the close relationship between JIT and people. The finding shows that the relationship between the use of JIT practices and manufacturing performance was not significant. However, there was a very strong relationship between JIT practices and infrastructure practices, which included quality management, workforce management, manufacturing strategy, organisational characteristics, and product design.

The necessity to focus on people issues extends beyond lean manufacturing and mass customisation. It extends to organisational attempts to enhance the level of agility in manufacturing. Several agility providers (practices, methods, tools, and techniques facilitating a capability for agility) also showed evidence of the need to scrutinise people issues. Through a survey that involved 1,000 companies and based on a total of 12 case studies (Zhang and Sharifi, 2000), they concluded that practices related to people and organisational issues are both important to manufacturers and highly effective in enhancing levels of agility in the organisation.

Despite the apparent importance of taking cultural imperatives into consideration when attempting to increase levels of mass customization, there has been little or no focus on issues that attempt to integrate the key dimensions of agile manufacturing systems. These dimensions relate to strategies, technology, people and systems. In a similar vein, there is a dearth of empirical evidence regarding the operational characteristics related to agile manufacturing.

There are key concerns related to a predominantly technology-focused approach in the adoption of agile manufacturing strategies. The ensuing argument was the need to have the right combination of strategies, culture, business practices, and technology for agile manufacturing.

The human factors correspondingly have a significant role in the implementation of agile manufacturing.

The relatively high emphasis placed on strategy, technology, and systems, and less emphasis placed on people as evidenced by the number of articles in the literature seemed to downplay the importance of human resource management in the context of new work or organisational culture.

Concept of Organisational Culture

Organisational culture represents how an organisation learns to adapt over time. It comprises common values that have been inherited and assumptions used over time by the organisation's members to analyse both their behaviour and that of others. It is the glue that holds an organisation together and is socially constructed.

Strategic leadership commitment transforms a benign culture into one that promotes competitiveness. Likewise, modifications in power base distribution and communication when a company implements lean manufacturing are also indicative of the management's commitment to improving organisational culture and as a result creating an agile organisation.

An essential feature of an agile organisation is that its members should be regarded as partners when implementing lean and agile manufacturing that enables the transition to successful delivery of intended results. This highlights the importance of innovative practices that enable organisations to exhibit high levels of adaptability, which are required to meet the ever-changing customer requirements.

Culture Dimensions

There are four dimensions in an organisation's culture.

They include

- mission and Vision
- adaptability
- involvement and
- consistency

This section attempts to show how these dimensions are manifested as attempts made to operationalise lean agility.

Denison (1990) defines cultural stability as the property of having high levels of agreement, consistency, and unity of purpose. Cultural flexibility is the property of having high-involvement and adaptable work practices. Consistency and mission are oriented towards cultural stability, whereas involvement and adaptability emphasise the organization's capacity for cultural flexibility and change.

Larger and more functionally organise manufacturing enterprises had to be oriented towards stability for them to engage in mass production.

However, as organisations began to engage in mass customisation, they had to become increasingly orientated towards a higher level of flexibility and agility and adopt a lean approach to manufacturing (Pine, 1993). This appears to suggest that from a cultural perspective, a shift towards cultural and organisational leadership flexibility is required for mass customisation.

People's Involvement and Empowerment

People's involvement and empowerment are critical for the factories of the future, implementing smart manufacturing and achieving a high-performing employee work culture. Organisations with high-involvement cultures have individuals who are empowered, whose capabilities are continually developed and who work in a team-oriented setting. This kind of culture is an important requirement for organisations that seek to enhance their level of mass customisation since the mass customisation process is dependent on how well an organisation can transform its leadership's role towards adopting a more empowering work style. Having a high-involvement culture is also a prerequisite to information sharing and developing trust among employees, which is another key requirement for mass customisation (manufacturing transistors) as evidenced by a study carried out by Liao et al. (2011). The progress that is experienced is synonymous with the employee's efforts and endeavours in any organisation, and this is possible with the dynamic involvement of those employees at the workplace.

The enhancement of agility through an increasing focus on improving employee involvement has been documented extensively in the literature. This suggests that unless the level of employee involvement is increased, the agility required to engage in mass customisation will be somewhat constrained.

The job function involvement model below summarises the investment of time and resources of the management functions to maintain the current established standard operating procedure (SOP) and to improve the current system for the better. Sustaining the current SOP is important, and this will ensure all expected results (e.g. in lead time, quality and cost). Improving the current work processes, new ideas, and breakthrough results will have to be the core responsibility of both the top leadership and the middle management team. This does not mean that the lower functional groups are not involved as they will be, but the leadership and managerial team must ensure they take total accountability and be personally involved to ensure that the various functional teams are delivering better results and performances continuously, which is the concept of 'every day a better day', the powerful, continuous improvement behaviour.

Figure 18: Job Function Involvement Model

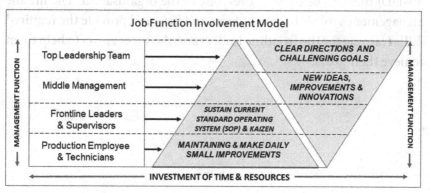

Source: Azlan Nithia, 2023

The maintenance of the current SOP must be strictly adhered to by the employees and the technicians who are involved in the daily production execution. This is their core job function that is expected of them all the time to ensure the smooth running of the daily operations and keep the production running per the quality and productivity targets. This group of employees will also participate in daily improvement activities, making small improvements daily and the small group activities (SGA), but their main function is to maintain and sustain the current SOP at the work place, any deviation must be elevated and approved.

The frontline leaders and supervisors have an important role in ensuring the employees maintain the current SOPs and deliver results that tally with all expected targets and goals. This functional group must likewise ensure that the real-time digital, visual performance displays are updated in a real-time manner and have the safety procedures and the 5S systems well managed. They must participate in the daily *gemba* kaizen activities (daily improvement) ensuring all equipment, machines, material flow, required resources, headcount, and technician support are available at the right time, in the right quantities all the time. Problem-solving will be this group's daily routine. For that reason, they are required to constantly identify the 3M waste (*muda*, *mura*, and *muri*) in the system to reduce and eliminate them. This requires a structured and simple problem-solving kaizen methodology (as you see in this book, A3-DD) that can be used by everyone in the organisation. The middle management's role is to give support and help and provide the required skills training for the frontline leaders and the supervisors to help them achieve their goals.

CHAPTER 7

Digital Technology and the Internet of Things (IoT)

Special Article by Azlan Nithia (2023)

The Impact of Digital Technology and the Internet of Things (IoT) on the Future of Manufacturing and Business

Abstract

Digital and internet of things (IoT) technology has opened a new paradigm that has fully changed the traditional ways of living to a higher-tech lifestyle. It is one of the most revolutionary and rapid advancements in recent years. The digital and IoT markets continue to grow steadily as the entire world experiences transformation. It facilitates the utilisation of embedded sensors, which collect operation data within an industry and send it to the artificial intelligence solution to find the means of optimising production. In manufacturing, the internet of things entails the use of advanced tools such as artificial intelligence and machine learning. These technologies improve manufacturing processes, allowing different firms to compete in the market effectively. The integration of IoT in manufacturing and other businesses will significantly impact the future of these sectors. Based on the research, the impact of IoT technology in the future of manufacturing and business will result in increased efficiency, reduced cost, increased customer satisfaction, increased connectivity, improved safety, and improvements in supply chain optimisation. This will increase the success of these firms in this sector while increasing the level of competition. The impact of this technology will also pose significant challenges that firms will need to prepare for, such as resistance, increased cost and the need for skilled personnel to operate and develop them.

Introduction

The IoT is a new paradigm that has fully changed traditional ways of living to a higher-tech lifestyle. Smart industries, smart transportation, energy savings, pollution control, smart homes, and smart cities are the changes that have resulted due to the IoT. Significant research investigations and studies have been carried out to enhance IoT technology (Kumar *et al.*, 2019). However, many issues and challenges must be addressed to ensure the realisation of the IoT potential. Precisely digital and IoT technology utilises the internet and smart devices to offer innovative solutions to different challenges that businesses, including the manufacturing industry, encounter globally.

According to Kumar *et al.* (2019), the IoT is a technological development that integrates sensors, intelligent devices, frameworks, and smart systems. It takes advantage of nanotechnology and quantum in terms of processing speed, sensing, and storage. This research paper focuses on the impact of digital and internet of things (IoT) technology on the future of manufacturing and business.

A Brief Overview of the Features of the Internet of Things

Being a connected technology, the internet of things has different features that allow it to be applied in different industries, including manufacturing and business organisations. According to Pedamkar (2023), the IoT device has different features that make it crucial to be applied in manufacturing and businesses.

Various features of this digital technological development include

Sensing

Humans can significantly analyse and comprehend circumstances based on previous experiences. In IoT cases, it is fitted with sensors that assist in gaining insights into a system in order to fully comprehend it. The sensors used in IoT include RFID, pressure, electrochemical, GPS, light sensors, pressure, and a gyroscope to collect data on a particular problem within the system (Pedamkar, 2023). The sensing feature is crucial for assessing system malfunctions.

Safety

It is also another crucial feature that the IoT system possesses. The connectivity feature passes sensitive information, particularly from the endpoint to the analytic layer in the IoT system. According to Pedamkar (2023), whenever designing for a company, it is critical to ensure that security measures, property safety, and firewalls are maintained to deter data manipulation and misuse. When the IoT systems are compromised, the entire system can fail.

Scalability

Designing the IoT with simple scalability is considered vital. Since the IoT is applied and implemented in different industries, ranging from the automation of large industries to smart homes, its scalability varies significantly (Borgia, 2014). Therefore, industries should design their IoT infrastructure based on their future and present engagement.

Active engagement

The IoT links different products, services, and cross-platform technologies together by creating active engagement. Cloud computing in the blockchain is utilised to create active engagement among the components of the IoT. According to Pedamkar (2023), 1% of the unstructured and 50% of the structured data are utilised to make important decisions within businesses. Therefore, according to the author, when designing ecosystems, carriers must take into consideration the future need to manipulate big data to fulfil the incremental needs of a business.

Connectivity

Connectivity is considered one of the main features a business or manufacturer could consider. Without having seamless communication among interrelated IoT ecosystem components such as data hubs, compute engines, and sensors. It is not easy to execute any proper business.

The IoT can be interconnected with various tools like microwaves, LiFi, and Wi-Fi to establish generic connectivity and maximise efficiency across IoT industries and other ecosystems. The connectedness of these systems helps ensure smooth operation while reducing any form of error. According to Borgia (2014), the success of any firm depends on how connected it is to improve efficiency, effectiveness, and higher performance.

How Digital and IoT Technology will Impact the Future of Manufacturing and Business?

The use and implementation of digital and the internet of things have transformed the way firms utilise data, communicate, and operate. It is projected that by 2025, more than 75 billion devices will be connected through the IoT (Kumar, 2023). In manufacturing and businesses, those changes will significantly come at a quicker pace, and a firm that was once slow to change will now digitalise at a higher speed. According to a study, the 2021 value of IoT in the manufacturing industry was about $62.1 billion. It will rise to $200.3 billion in 2030 (Kumar, 2023). The adoption of IoT and other digital technologies will have positive impacts on the manufacturing industries in various ways as discussed below.

Increased Efficiency

In an environment where efficiency is considered crucial, manufacturing industries are in a race to fully outshine their competition. The use of digital and internet of things technology will enable manufacturers to comprehensively monitor their operations in real time, automate processes, and predict machine failures. This will enhance streamlined operations that will lead to lower costs, increased efficiency, and higher octane results, therefore increasing the levels of the firm's competition. Kumar (2023) notes that IoT will improve efficiency in manufacturing by ensuring real-time maintenance and monitoring. For instance, the IoT sensor can ensure effective monitoring of the machinery's performance in real time, predicting possible errors and breakdowns.

This predictive maintenance will help improve overall equipment effectiveness and minimise costly downtime. Moreover, according to the author, digital and IoT technologies will help improve supply chain optimisation. For example, IoT allows for precise monitoring and tracking of all materials and finished products as they move via the supply chain. This will result in decreased waste, improved inventory management, and a more efficient and responsive supply chain. Löffler and Tschiesner (2013) note that the implementation of IoT in manufacturing will trigger a paradigm shift in manufacturing. This shift will positively impact the manufacturing value chain and classic production, increasing efficiency.

Also according to a systematic review conducted by Kalsoom et al. (2021), the implementation of IoT technology in manufacturing and business would be vital for improving productivity and operational efficiency. The manufacturing firms will be able to combine different devices like RFID and sensors, wireless networking, and barcoding to realise enhanced visibility of various activities, particularly within the manufacturing and business facilities. By doing so, these firms will be able to achieve higher productivity and operational efficiency (Kalsoom et al., 2021). Efficiency and increased production are key for any manufacturing and business organisation to succeed and ensure effectiveness in their levels of competition.

Improved Safety

Manufacturing accounts for about 15% of the workplace's fugitive emission of toxic fluids, illnesses, injuries, improper use of PPEs, and excessive radiation triggered by faulty equipment (Christiansen, n.d.). Safety is another vital concern for manufacturers; they heavily invest in security systems that are key for controlling access, suppressing and detecting fire, tracking assets, and preventing theft of assets (Christiansen, n.d.). The digital and internet of things technologies will provide game-changing opportunities for the manufacturing industries and assist them in overcoming prominent challenges and automating processes.

Various digital and IoT technologies will help improve workplace security and safety. For instance, its implementation will help improve asset maintenance within the manufacturing industry. The implementation of digital technology and IoT sensors will transform the quality and scale of maintenance in manufacturing facilities (Christiansen, n.d.). Fixing the production equipment with sensors that can monitor the real-time performance of the equipment will assist in detecting and predicting any underlying failure or defect.

Moreover, with the manufacturing industries and businesses struggling to maintain the visibility of all product assets, including protecting them from theft, the use of digital and IoT technology will assist in centralising the operations and management of such assets (Santhosh et al., 2020). Sufficient and accurate data from the IoT systems will assist manufacturers in the near future to lower the risk assessment turnaround (Kalsoom et al., 2021). The manufacturers usually utilise the insights from the assessment to strengthen safety compliance and increase the reliability of the external and internal security systems.

Nevertheless, security and safety will be improved through monitoring employee safety and wellness compliance. This is because fatigue is among the leading triggers of workplace incidents and injuries in the manufacturing industry. It is essential to ensure that the technicians and operators within manufacturing industries are fit physically to interpret real-time data related to products (Kalsoom et al., 2021). The IoT sensors that are fitted in wearables and personal protective equipment can detect hazards and monitor employee wellness.

Smart sensors can ensure the effective detection of hazardous environments, internal machine defects, fires, and smoke. It must be noted that the utilization of the wearables will assist leaders in the manufacturing industry to keep monitoring things like the surrounding noise levels and employees' posture; this will, in turn, help improve performance and work conditions (Kalsoom et al., 2021). The wearables can also notify employees when they are not carefully following the safety procedures in the workplace so that they can correct their actions to ensure their actions align with the workplace regulations, hence improving their safety on the job.

In general, the IoT will assist manufacturers across the globe to improve their security and safety standards while enhancing their compliance with statutory regulations. Manufacturers will also need to address the most pressing security and safety challenges prior to implementing and selecting a suitable IoT (Moisescu et al., 2010). They will also need to secure digital systems to safeguard the accuracy and integrity of the data gathered and relayed through these systems.

Reduced Costs

The way in which digital and internet of things technology lower costs differs significantly across different industries. In manufacturing and business organisations, cost reduction has been an important aspect of ensuring firms increase their revenues (Bither, 2020). Because this is necessary for the success of a firm, the implementation of digital and IoT technology within these areas will be of greater importance. Thus its implementation will help firms save money by reducing maintenance costs. The IoT will also help bridge the gap between the executives and the floor workers and identify any bottlenecks in production. Predictive maintenance, which arises due to the use of IoT, will assist manufacturing and businesses in reducing costs by lowering customer dissatisfaction and employee churn due to quality concerns, remote monitoring with real-time alerts, and replacing parts, especially when they need to be replaced whenever they demonstrate signs of malfunction or are based on historical data. This will assist in reducing costs (Bither, 2020). Additionally, the use of automation will be critical for businesses and manufacturing industries. For instance, by linking devices and machines to the internet, the manufacturers will be able to remotely control and monitor them, lowering the need for human intervention. This reduction in human intervention will result in a significant reduction in the cost of labour and increase accuracy and efficiency, thus lowering the cost.

Supply Chain Optimisation

The supply chain is considered one of the most critical operations of any business, including manufacturing performance. An exceptional supply chain is considered key to the growth of any business. It fully allows the stakeholders in the manufacturing sector to be fast-paced. A simple integration of the internet of things into the supply chain can be considered essential in transforming the manufacturing industry supply chain. According to Gerlée (2023), there are various ways in which the digital and the internet of things impact manufacturing and business organisations in optimising the supply chain.

The first is through improving the supply chain documentation. Documentation during each supply chain stage will assist in recording system performance to examine any form of delay, gap, or inadequate behaviour. The digital and IoT technologies will help in optimising the whole supply chain process, from procurement all the way to logistic delivery.

Moreover, the IoT will help enhance quality control since the sensors can significantly ensure that the devices are perfect even after transit. The quality control through the sensors will also ensure the transit process is more safe and secure (Kiel et al., 2017). This will, in turn, assist in improving the overall supply chain process within the manufacturing and business sectors.

The IoT will also impact manufacturing by ensuring proper inventory and warehouse management. This is because, just like the machine, the stock and the inventory of the raw materials are considered vital in ensuring smooth chains of supply. According to Kiel et al. (2010), IoT solutions and sensors like printed codes and radio-frequency identification can assist in tracking inventory and ensure automatic raw material procurement whenever needed. This will assist in reducing overstocking and stockouts. The utilisation of IoT in the supply chain will help improve the manufacturing industry and other businesses' monitoring and real-time tracking. The implementation of the tracking system via the IoT will assist in locating the activities of the assets and the product. The outdoor and indoor tracking will greatly assist businesses and manufacturers in optimising and supervising supply chain performance.

According to Pal (2023), the implementation of digital and IoT technologies will simplify supply chain management in manufacturing and businesses. This improvement will assist in increasing efficiency in all aspects of the supply chain process. The presence of the sensors will ensure the overall processes are improved, thus preventing any aspects of delays and other forms of supply chain disruptions. The use of radio frequency identification will also be a vital aspect of the IoT that will significantly assist in improving supply chain management by improving traceability, inventory management, and operational efficiency (Rejeb et al., 2020).

Improved Customer Satisfaction

Presently industrial customers demand customer experience and service-oriented buying. They also expect constant and prompt customer service availability and interactive and mobile access to data and products. Within manufacturing and business, IoT technologies have assisted in providing detailed insights into inventories and supply chains (Soldatos et al., 2022). This means that through this form of technology, the goods and services will be able to reach the customers faster, and precisely how the goods and services from businesses and manufacturing ordered will arrive and where and when the supply chain was interrupted. This will improve the customer's efficiency and satisfaction.

According to Günes (2020), firms and other business organisations that normally take the aspect of customer satisfaction and efficiency seriously derive multiple positive impacts from long-term customer retention, higher sales, willingness to pay, and higher recommendation rates. Therefore, for manufacturing and business organisations, implementing the use of digital and IoT technology will positively impact the levels of customer satisfaction as Günes (2020) reveals.

Similarly according to Ahmed et al. (2022), customer satisfaction is vital for realising profitability through minimising operation costs and profits. Customer satisfaction is the result of the evaluation processes since it arises from the feedback the client provides; therefore, it is significantly impacted by the satisfaction levels of the client. Customers will tend to be more satisfied if their original expectations exceed or are met. The implementation of IoT in manufacturing and business will help improve customer experience and satisfaction. This is because the impacts that the internet of things has on customers are direct and significantly beneficial. The improvement in the supply chain by reducing the higher rates of downtime and supply chain disruption through the use of the IoT will greatly help increase customer satisfaction and experience (Yang et al., 2016).

Increased Interconnectivity of Systems

Interconnecting systems within an organisation, particularly manufacturing and business, is vital for ensuring better performance. The IoT in manufacturing usually focuses on linking all individual machines and assets into one system (Soldatos et al., 2022). The overall goal of doing so is to establish easily manageable systems that offer real-time performance monitoring across distinct production plants. This means that IoT technology reduces the gap between different assets by fully tracking performance and finding new means of improving efficiency. The interconnectedness of the systems via the IoT generates data performance that increasingly allows artificial intelligence to learn how different assets perform. The interconnectedness of systems is crucial for instantaneous monitoring within manufacturing, making it easier for managers to make better decisions, monitor machine performance, and access real-time operational data. Furthermore, this will increase the operational efficiency of manufacturing and other business operations.

Improved Decision-Making

The pervasive sensing ability of the internet of things usually gives rise to the generation of diverse and huge data. The manufacturing company will utilise the big data to assist in making optimal decisions (Yang et al., 2018). Every business requires effective decision-making, which is key to the success of manufacturing and business organisations. Essentially, cloud computing integrated with IoT technology will facilitate firms in decision-making. This will assist manufacturing to better capture business opportunities, ensuring they positively respond to any uncertainty and adapt to changes.

However, despite the positive impacts of digital and IoT technology, the increased implementation and use of the technology will pose possible threats to future manufacturing and business organisations. For instance, according to Tomic (2017), every technology has its setbacks. For instance, the increased setbacks of the existing infrastructure, which in many cases require many changes to ensure the internet of things works effectively.

The concern is that existing structures in many manufacturing sectors and business organisations are not autonomous or interconnected, forcing manufacturing to be redesigned. For example, the digital infrastructures for the use of RFID tags, motion detectors, and several sensors would need to be redesigned to ensure the technology works effectively with the organisation's current digital infrastructures. Redesigning the entire manufacturing process will require significant costs, which most businesses and manufacturing firms are not ready to incur (Tomic, 2017). Nevertheless, as the article author notes, the increased investment in the redesign and integration of sensors will make it difficult for manufacturing industries to sustain, making it difficult to implement the IoT.

In a similar study, it was found that the implementation of IoT had significant challenges for manufacturing and business organisations. For instance, immature technology and the uncertain return on investment presented a notable barrier for potential adopters. Importantly, the same report of the study revealed that cyberattacks and data security were additional burdens that manufacturing firms would incur due to the increased data sharing and interconnectivity (Tomic, 2017). This is because the majority of the data privacy concerns presented by the IoT are not different from those triggered by existing digital technologies. The IoT is connecting physical resources with digital operations, where the data controls the most physical part of the system. This demonstrates that a security breach may damage both the physical production and the data.

Existing knowledge and labour skills are other challenges that the implementation of the IoT will have in the future of business and manufacturing. This is because having the most efficient and effective system would significantly require adequately skilled manpower who can monitor it, comprehend how it works, and conduct further development on the system. In the future, there will be a likelihood of a decline in traditional jobs with the increase in demand for competent people to design IoT systems and manage the program effectively (Tomic, 2017). Findings from the same study revealed that the impact of increased levels of digital technology innovation will cause the majority of the manufacturing industries' jobs to become obsolete. Increased production caused by the IoT will result in the rationalization of the input factors. Many employees will be laid off since the huge production output generated in a shorter period by these smart technologies will render human labour ineffective and in less demand.

Conclusion

The impacts of digital and the internet of things in the business sector and manufacturing are diverse. Leaders in the manufacturing sector and other business organisations need to comprehend the impacts and possible threats that the new technology will pose to the manufacturing sector. As much of the research identified above suggests, the benefits of digital and internet of things technology will open a new chapter for manufacturing firms by ensuring that they integrate the technology to improve safety, increase customer satisfaction, improve efficiency, enhance supply chain optimisation, and improve the decision-making process. Integrating the system will also improve production by improving efficiency and reducing human labour. The success of manufacturing firms lies in the integration of digital and IoT. However, leaders of these firms need to be aware of the possible challenges posed by the integration of these technologies, such as barriers to embracing the technology, cost implications, and quality manpower to manage and operate these systems. Addressing these challenges will help enhance the success of manufacturing industries and other businesses that embrace technology. Furthermore, it will help manufacturing firms to have a competitive edge over their competitors since they produce products quicker and in an efficient and effective way compared to the use of human labour.

We must harness advancing digital technologies to create a highly productive and more equitable future in all business and manufacturing industries.

The Future of GPS for the Indoor Positioning Industry

(An example of GPS technology in healthcare)

The healthcare application within the indoor positioning market presents a compelling solution to the challenges posed by an ageing population and the early detection of cognitive decline. It is essential to discuss the most intriguing positioning market, how the product offering would look for a new market entering the indoor positioning market and describe the three possible markets where indoor positioning is a key solution.

The healthcare application within the indoor positioning market is most and particularly intriguing. With the ageing population and the challenge of early detection of cognitive decline, integrating indoor positioning technology offers a promising solution. The ability to automatically measure mental health indicators, such as gait speed, at home through efficient means like small GPS trackers can significantly assist in proactive healthcare strategies. It addresses the pressing problem of relating to decline in older people and streamlines the observation and checking process, thereby lowering the burden of human resources. The possible impact on enhancing the quality of life for elderly individuals and making the intervening process much earlier makes the sector of the indoor positioning market both compelling and essential.

The ideal product offering would comprise a comprehensive and user-friendly solution for a new company in the indoor positioning market focusing on healthcare applications. The product could include a suite of small, discreet GPS trackers equipped with advanced sensors tailored for elderly individuals and those with cognitive impairments. The trackers would seamlessly integrate with a user-friendly mobile app, providing real-time location tracking, fall detection, and continuous monitoring of vital parameters. The product could offer data analytics tools for trend analysis and early detection of cognitive decline based on gait speed and movement patterns to cater specifically to healthcare professionals. The emphasis on simplicity, accuracy, and adaptability would set the product apart, addressing the critical need for effective and non-intrusive solutions in the growing healthcare-oriented indoor positioning market.

The three possible markets where indoor positioning is a key solution include the retail sector, healthcare domain, and contact tracing market. The retail industry influences indoor positioning for personalised marketing and increases customer experiences. Furthermore, the healthcare domain uses indoor positioning for elderly safekeeping, tracking patients' motion and movement, and providing for early detection of cognitive decline with a continuous market known for its substantial growth. In addition, the contact tracing market, driven by the urgency of public health concerns such as COVID-19, utilises indoor positioning through technologies like Bluetooth, representing a rapidly evolving sector with substantial investment. These markets collectively project a multibillion-dollar industry, showcasing indoor positioning solutions' versatility and expansive reach. The estimate of the size of the market range is several billion dollars. The increasing population and growing demand for non-intrusive monitoring solutions in healthcare provide substantial market potential.

The favourite alternative centres on healthcare applications, explicitly designing a product for checking, observing, and detecting cognitive reduction in older people. The location technology would integrate small GPS trackers with inertial sensors, capturing movement and gait speed. Additionally, ambient sensors for environmental context could enhance data interpretation.

A combination of sensor fusion algorithms, machine learning models, and edge computing would be necessary to extract location information. These algorithms would process data from GPS, inertial sensors, and environmental sensors to derive accurate location details. The computing power, likely a combination of on-device processing and cloud-based analysis, would ensure real-time insights for healthcare professionals and caregivers.

In conclusion, the healthcare application in the indoor positioning market holds significant promise, particularly in addressing the challenges of an ageing population and early detection of cognitive decline. Integrating technology like small GPS trackers for automatically measuring cognitive health indicators, such as gait speed, at home offers proactive healthcare strategies.

References

Ahmed, M., Abdou, M. Y. K., and Elnagar, A. M. (2022) 'The impact of implementing the Internet of Things (IoT) on customer satisfaction: evidence from Egypt', *Journal of Association of Arab Universities for Tourism and Hospitality*, 22(2), pp. 365–380.

Bither, B. (2020) '5 ways industrial IoT reduces costs for manufacturers'. Retrieved from: https://www.machinemetrics.com/blog/industrial-iot-reduces-costs.

Borgia, E. (2014) 'The Internet of Things vision: Key features, applications and open issues', *Computer Communications*, 54, pp. 1–31.

Christiansen, B. (n.d.) 'How does IoT help rise above safety and security challenges in manufacturing'. Retrieved from: https://reliabilityweb. com/how-does-iot-help-rise-above-safety-and-security-challenges-in-manufacturing.

Gerlée, K. (2023) 'The Benefits of Implementing IoT in Supply Chains'. Retrieved from: https://freeeway.com/the-benefits-of -implementing-iot-in-supply-chains/

Günes, U. (2020) 'IoT takes customer relations to new level'. Retrieved from: https://iot.telekom.com/en/blog/iot-boosts-customer-relations.

Kalsoom, T., Ahmed, S., Rafi-ul-Shan, P. M., Azmat, M., Akhtar, P., Pervez, Z., . . . and Ur-Rehman, M. (2021) 'Impact of IoT on Manufacturing Industry 4.0: A new triangular systematic review', *Sustainability*, 13(22), p. 12506.

Kumar, S., Tiwari, P., and Zymbler, M. (2019) 'Internet of Things is a revolutionary approach for future technology enhancement: a review', *Journal of Big Data*, 6(1), pp. 1–21.

Kumar, M. (2023) 'Increase efficiency of manufacturing operations with IoT solutions'. Retrieved from: https://www.datasciencecentral.com/ increase-efficiency-of-manufacturing-operations-with-iot-solutions/.

Löffler, M., and Tschiesner, A. (2013) 'The Internet of Things and the future of manufacturing'. Retrieved from: https://www. mckinsey.com/capabilities/mckinsey-digital/our-insights/ the-internet-of-things-and-the-future-of-manufacturing.

Kiel, D., Arnold, C., and Voigt, K. I. (2017) 'The influence of the Industrial Internet of Things on business models of established manufacturing companies—A business level perspective', *Technovation*, 68, pp. 4–19.

Moisescu, M. A., Dumitrache, I., Caramihai, S. I., Stanescu, A. M., and Sacala, I. S. (2010). 'The future of knowledge in manufacturing systems in the future era of Internet of things', *IFAC Proceedings Volumes*, 43(17), pp. 215–220.

Pal, K. (2023) 'Internet of Things Impact on Supply Chain Management'. *Procedia Computer Science*, 220, pp. 478–485.

Pedamkar, P. (2023) 'IoT features'. Retrieved from: https://www.educba.com/iot-features/.

Rejeb, A., Simske, S., Rejeb, K., Treiblmaier, H., and Zailani, S. (2020) 'Internet of Things research in supply chain management and logistics: A bibliometric analysis'. *Internet of Things*, 12, p. 100318.

Santhosh, N., Srinivsan, M., and Ragupathy, K. (2020, February) 'Internet of Things (IoT) in smart manufacturing'. In *IOP Conference Series: Materials Science and Engineering*, Vol. 764, No. 1, p. 012025. IOP Publishing.

Soldatos, J., Gusmeroli, S., Malo, P., and Di Orio, G. (2022) 'Internet of things applications in future manufacturing'. In *Digitising the Industry Internet of Things Connecting the Physical, Digital and VirtualWorlds*, pp. 153–183. River Publishers.

Tomic, D. (2017) 'The benefits and challenges with the implementation of the Internet of Things (IoT) manufacturing'.

Yang, C., Shen, W., and Wang, X. (2016, May) 'Applications of Internet of Things in manufacturing'. In *2016 IEEE 20th International Conference on Computer Supported Cooperative Work in Design (CSCWD)*, pp. 670–675. IEEE.

Yang, C., Shen, W., and Wang, X. (2018) 'The Internet of things in manufacturing: Key issues and potential applications', *IEEE Systems, Man, and Cybernetics Magazine*, 4(1), pp. 6–15.

CHAPTER 8

Smart Manufacturing–
Factory of the Future

The Inevitable Manufacturing Transition

To reimagine the manufacturing industry achieving global competitiveness through the successful adoption of greater digital technology of a connected factory to increase efficiency, speed, and agility requires the ability to be able to rapidly transition into the new smart manufacturing paradigm and be quick in responding to the demanding market changes in a digitally connected marketplace. This will sustain higher levels of customer competitiveness.

With the advent of smart manufacturing in the factory of the future, there is an urgent need for industries and organisations to transform their manufacturing operations, supply chain, and people's culture. This can be done by successfully implementing the requirements of the factory of the future that will better position the industry to improve customer responsiveness, resulting in becoming a globally competitive supplier. The successful implementation will require the industry to be ready with a strong manufacturing foundation and operational excellence. This aids the implementation of smart manufacturing, and IoT (internet of things) to achieve the status of a connected factory of the future; achieved efficiently by the continuous growth of business competitiveness and improvement of the company's profitability.

The digital and cyber-physical technologies that are impacting the industries with smart manufacturing must be at the forefront of all the manufacturing industries. The industries will continue to be influenced by Industrial Revolution's technological advancements in discrete and process manufacturing, supply chain, marketing, and consumer interface. The industries have already experienced, over the decades, various industrial revolutions since the first invention of mechanised machines, followed by conveyor system production, mass production, and the emergence of digital technologies that started challenging and changing manufacturing platforms. Manufacturing is evolving into a sophisticated cyber-physical system, connected end-to-end supply chain (or value chain) coupled with the application of industrial IoT, and connected intelligence in the related industrial aspects of manufacturing, production, and logistics.

The Industrial Revolution 4.0 was initially conceived around 2015 within the context of manufacturing applications, but this quickly expanded to all industries around the globe. The discussion of Industrial Revolution 5.0 had already started in the year 2018. The duration between the industrial revolutions is getting shorter. For example, the dissimilar period between the first and second revolutions as compared to the third and fourth revolutions, Industrial Revolution 5.0 is already concurrently happening with the fourth revolution. Future industrial revolutions will continue to develop in parallel and concurrently within shorter durations.

You must understand the changing technological environment and how necessary is digital technology with connected processes for your organisation and the business! This must be your starting point before diving deeper into any digital technology!

It is important to understand the changing market and business environment. Constantly, from time to time, businesses and organisations must innovate, review the level of digitalisation required, and process automation needed to cope with

- competitors and competition
- new products and changing product technology
- changing government policies
- changing consumer preferences

To understand the benefits and impacts of the rapid industrial transition into smart manufacturing, it is essential to see the full value chain. This includes suppliers of the materials and components needed for internal manufacturing processes, the connected end-to-end digital supply chain, and the final destination connected digitally to the end consumer.

Personalised products, mixed model production, increased variety, and quick response to consumer demands are some of the goals that can only be achieved with the connected smart, customer-centric factory of the future. These industries must respond to an increasing sense of demanding customers who simply value speed, cost, agility, and innovative value-added services.

In the end, all industries will remain as businesses, albeit smartly connected businesses in a digitally connected world. These smart businesses will have the innovative twist of innovation and transformational digital technologies of business models and processes that will increase profit and decrease product costs. This will enhance the consumer experience, optimise consumers' loyalty through lifetime value, increasing the global market with innovative growth. The customers will invariably remain relevant and responsive to any market digital disruptions.

We will always find in a digital industrial revolution, a matured technological ecosystem, new disruptions to current technologies, new advancements, and advancing capabilities with quantum leaps. Changes to the digitally connected world will be happening faster than ever.

The Industrial Revolution

In the last two hundred years, we have seen quantum leaps in technological innovations, continuous advancement of technology, and industrialisation. Between the mid-1700s and early 1900s, there were many major inventions which can be classified as Industrial Revolution 1 and 2 (IR 1.0 and IR 2.0). These two industrial revolutions nurtured many inventions that were the basic and foundational needs for people, examples being trains, steam engine ships, cars, planes, telephones, electric bulbs, radios, and tube television. Then the Industrial Revolutions IR3.0 and IR4.0 followed from the late 1900s into 2000s. This was the era of digital and advanced technology like fax machines, mobile phones, the internet, Wi-Fi, computers, electronic calculators, electric trains, high-speed trains, faster planes, digital televisions, high-performance semiconductors, programmable logic controllers, computer-controlled machines, Google search engines, Facebook, WhatsApp, and many more. This resulted in connecting the world and people digitally and faster. This increased the contemporary perception of the world being a small place and as a result, rapidly advanced and connected the digital global trade.

The Industrial Revolution 4.0 which started in the early twenty-first century has advanced technologies developed to connect machines to people so that both can work collaboratively. For example, advanced robots can be operated remotely thousands of kilometres away, and with very high precision, the internet of things (IoT) connects consumers, product designers, and manufacturers in real time, working on common platforms, increasing the speed to market.

The generations of the Industrial Revolution can be briefly explained below:

1. **Industrial Revolution 1.0**
   IR 1.0 started with the era of mechanical systems and systems that used steam to power steam engines and other mechanical movements. This is also when the

mechanical loom was invented and later the factories to manufacture cloth. This further stimulated the cotton and textile industry, assuring the beginning of the first industrial revolution.

2. **Industrial Revolution 2.0**

   IR 2.0 saw many industrial innovations and inventions that changed the industrial landscape rapidly. This revolution saw advancement in electricity production and supply, mass production systems, conveyors and belts run by electric motors, faster trains, telephones, motorcycles, cars, aeroplanes, steam power ships, and televisions to name a few inventions. Numerous mechanically automated machines were used in the manufacturing industries to achieve higher levels of productivity output with better quality. Other industries like farming and construction started using mechanically operated equipment to increase their output and accordingly lower the cost of production. The Ford Model T motorcar was introduced in 1913, which rolled out of the Ford factory's assembly lines. This marked the beginning of high-volume mass production. These industrial developments continued to evolve and progressed rapidly throughout the world, and later mechanical automations were used to further increase production outputs, improve quality, and reduce cost.

3. **Industrial Revolution 3.0**

   The application of computers, IT systems, electronic, high-performance semiconductors, programmable logic controllers, multi-function IC chips (dual-core chips), high-performance, micro-sized microprocessors, advancement in mobile phones, Facebook, Google search engines, digital libraries, WhatsApp, WeChat, internet banking, wireless technology, mobile networks became faster from 1G to 4G (fourth generation mobile network). The application of advanced automation to reduce lead time and improve quality, and customer responsiveness became a key competitive advantage with the introduction of robots in manufacturing, fibre optics, high-speed electric trains, electric cars, and many more devices.

The globalisation of businesses' advanced digital supply chain systems and internet marketing changed the way we forecast customer requirements, manufacture products, and market products and services globally. These advancements and high-tech industrialisations have made the world a small place. Marketing products digitally and globally has evolved in almost every industry.

4. **Industrial Revolution 4.0**

The continuation of Industrial Revolution 3.0 successes evolved further into Industrial Revolution 4.0. The convergence towards digital, cyber-physical systems started creating numerous disruptive digital and intelligent processes where technological innovations disrupt the manufacturing value chain. The operational excellence by leveraging disruptive technologies to develop new business models, advanced analytics, additive manufacturing (3D printing), advanced robotics, industrial IoT (internet of things), and digitally connected factories. It presented an immense opportunity for intelligent processes and systems to achieve better quality, higher efficiency, increased speed to market, product co-creation, increased customer responsiveness, and a flexible and responsive production system.

The Industrial Revolution 4.0 has made all industries scramble to understand what digital cyber-physical systems and advanced automation can do for them and how the systems could change their company's competitiveness in the marketplace. It has become overwhelming for many businesses to invest in the capital expenses required to implement digital connectivity, IoT, robotics, and automation. A critical point for decision for any industry has arrived which is either the organisation transforms or embraces IR4.0 or fails to adopt the conceptualised digitally connected factory and gets eliminated.

A digitally connected factory requires the full implementation of smart manufacturing technologies that fully adopt automation and human-machine interface to ensure full traceability from customer order to delivery. There is an urgency to understand that the cost of manual labour will continue to rise while the cost of automation will continue to go down.

The development to upgrade the 4G mobile digital network has become urgent as the remote robots are required to operate with extremely high precision to accurately perform the various activities in real-time on various jobs from a remote station located hundreds or thousands of kilometres away. The emergence of a 5G digital network that operates from up to fifty to one hundred times faster than 4G is critical to successfully implement functions like using remote robots to perform open-heart or even brain surgeries remotely without human surgeons and many other sophisticated remote-controlled equipment.

This fourth revolution has reset the people's readiness backwards. The concern now is how ready are they to learn fast, adopt, and implement new digital technologies flawlessly. The teaching methods and the subjects taught in schools, colleges, and universities have to be revised to include advanced digital technologies, robotics, and industrial IoT among many other related subjects. It requires a high level of ability to continuously learn new methods, adapt quickly to changes, and respond to customers quickly, able to work on multiple processes and have the ability to respond to real-time digital data.

In a nutshell, the concept of Industry in IR 4.0 encapsulates the development of the manufacturing sector from a laborious mass production model to an integrated, seamless automated connected factory, the factory of the future and a digitally connected end-to-end supply chain system from consumer order to delivery. There is a greater need for more diverse and complex skill sets to improve productivity in high-mix quota (volume) production environments, market developments, and consumer preferences will push companies to develop more innovative solutions to meet the evolving demands of the present and future consumer needs.

5. **The Industrial Revolution 5.0**

   In the new era of modern digital intensive manufacturing, artificial intelligence (AI) and collaborative robots will be at the forefront of this revolution (IR 5.0). There will be a greater application of human-robot collaboration (with infinite possibilities) in manufacturing processes and many other industries including deskilling and remote workstations in the medical industry. The co-creation of products with customers with customer obsession, mass customisation, agility (people and processes), and customer experience will become a norm for every industry. The smart manufacturing application in the manufacturing industries will be the standard and a critical prerequisite for competing in a digitally connected marketplace while achieving environment-friendly zero-waste production and sustainability. A much more advanced human-machine connected interface system will continue to be developed and applied in many different industries.

The future of driverless cars looks very promising. As technology continues to advance, autonomous vehicles will become more and more prevalent on our roads in the future. They have great potential to revolutionise the transportation industry by better safety, reducing congestion with smart traffic flow, and increasing accessibility. In the coming years, we can expect to see increased adoption of driverless cars by both individuals and businesses. Companies like Tesla and Uber for example are investing heavily in autonomous vehicle technology. As technology, safety, and regulations continue to improve, we may see fully self-driving cars or vehicles becoming a reality and common.

High-quality 3D printing will be the industrial production reality, especially with various advancements in material sciences and special material technology. The additive manufacturing processes will eliminate the time-consuming manufacturing processes. The medical industry will benefit greatly from additive 3D printing advancements, for example, the printing of very fine layers of reconstructive bone parts at high levels of precision coupled with the ability to construct very complex bone contours using super lightweight materials.

The new product introduction process from product ideation to product availability in the market lead time will be extremely short with the use of additive 3D printing. It will eliminate all the manufacturing processes like tooling and equipment. The time to produce the product will be equal to the time to print the parts. The high-speed 3D part printers will print a variety of different parts, as well as personalised preferences offered by suppliers as and when required with high precision, at high quality, personalised, and at lower cost. This will change the landscape of the manufacturing industry globally.

The 5G digital network that operates up to fifty to one hundred times faster than 4G will further be enhanced to even faster digital connectivity. This will increase the number of remote human-like robots, for example, used to perform surgeries on the human body more widely in the global, borderless medical industry. This will reduce the need for highly skilled surgeons because these human-like remote robots can perform surgeries very precisely and work continuously for twenty-four hours every day. This advanced digital network technology will even be used to remotely navigate aeroplanes, cars, drones and self-driving vehicles.

The Industrial Revolution will continue to evolve and improve digital connectivity and advanced technologies by creating a seamless global market competition and complete transparency of every single activity happening in the industrial world, especially the consumer's preferences or demand patterns. It will require strong organisational agility and people learning fitness to be able to continuously learn, adopt, and implement newer industrial digital advancements rapidly as and when they become available. The speed of adoption does matter and will be significant for flawless quick implementation to capture and apply the new business competitive advantage thus sustaining the continuous profitability of the company. Every time a new technology is adopted, it creates a new value for the organisation. The value chain must also be connected with this revolution to increase customer responsiveness. This must be both the organisation's ethos and end goal.

The future of the industry is about the survival of the fittest. The fittest are those organisations that are quick to learn and quick to implement new technologies flawlessly and to strengthen and grow the business rapidly and globally.

Organisations must compete like a fish in the ocean (globally) and not like a fish in a pond (domestic).

Steps to Become a Factory of the Future

The capital investment required to implement the technologies, automation, robots, and connected factories is very high. It is extremely important for the company to ensure it has assessed its gaps, upgraded its foundation so that it can successfully apply the advanced technologies and established clear evidence for the return of investments (more agile manufacturing, shorter lead times, faster model-to-model changeovers and lower product cost).

In summary, the investment in advanced technologies is to further help the company achieve a higher level of customer responsiveness and be able to deliver products faster at lower cost.

The end in mind must be very clear where investments are strategically connected to the final expected business results.

There are four critical steps (or the AURI model) required of any company to step up its technological capabilities to become a digitally connected factory of the future with a connected world-class supply chain system. The AURI model stands for assessment, upgrade, readiness, and implementation.

The Four Steps (AURI Model) to Achieve Factory of the Future (FOF)

- **Step 1: Assessment**
  This is an important first step to discovering the current stage of the company and its manufacturing excellence gaps. This assessment is key to the establishment of the amount of work required in both the manufacturing processes and the people readiness gap.

  The people assessment takes into consideration the ability and the capability of the current people in the organisation, to be able to adapt to new technologies, have the capacity to continuously learn new ideas, and have the ability to be able to implement new technologies. This requires an agile organisation that is able to adapt to changes, which includes the leaders of the company.

Figure 19: Foundational Steps to Become a Factory of the Future (AURI Model)

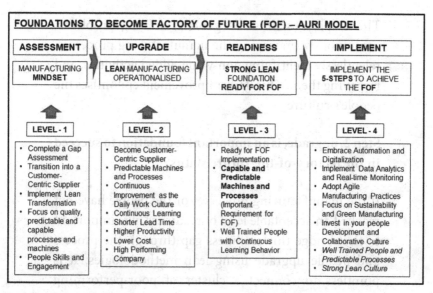

Source: Azlan Nithia, 2023

- **Step 2: Upgrade**
Based on the assessment of the current stage, the company has then to be upgraded (capability of the processes and its people culture) into a readiness stage, that is, to be a customer-responsive, customer-centric, high-performance organisation fully embracing and deploying all the lean manufacturing principles.

A lean organisation is an organisation that has operationalised capabilities focused on predictable processes and machines (with good overall equipment effectiveness—OEE) producing good quality parts, demonstrating high productivity, and continuously improving all these further to make the organisation's productivity better and better. This must be a relentless journey of continuous improvement activities. Achieving a high level of manufacturing processes and machine efficiencies requires employees who can actualise and constantly improve it.

The employees must be good at solving problems, continuously improving the manufacturing processes, relentlessly identifying non-value-added activities, and removing them. This daily improvement epitomises the people's culture.

Step 2, upgrade, is an important foundation to succeed in the factory-of-the-future status.

If the manufacturing processes or the people have not reached an acceptable level of readiness, it is important to first bridge this readiness gap through the process and people upgrade using lean methodologies. It is pointless to connect a cluster of poor-performing machines to a central monitoring system and not have capable employees solve the problems rapidly. This will only cripple the company because it is not ready to embrace the higher levels of advanced technologies. This will result in wasted capital investments and may even negatively impact the business.

Therefore, having capable machines and capable employees is the prerequisite to embarking on smart manufacturing—the connected factory of the future. Once the organisation has implemented a strong lean manufacturing foundation, it is ready to upgrade its manufacturing processes and supply chain system with smart manufacturing technologies.

- **Step 3: Readiness**
  This is the step closer to becoming the factory of the future. Step 3 is about ensuring all the operational eco-systems are ready. This step will ensure that the investments are strategically invested in the right advanced digital technologies, automation, and IoT. At this readiness stage, all the processes and machines have been upgraded to a predictable and capable level to produce good quality parts all the time efficiently. The people are engaged in daily, continuous improvement activities, continuously learning new things to apply daily problem-solving and good 5S.

  It is very important to ensure the investment and the types of digital and automation technologies will improve the company's customer responsiveness, reduce delivery lead time, lower product cost, and position the company on a sustainable growth path and become a globally competitive company. If it does not deliver on those benefits, then the return on investment (ROI) may not be achieved as planned.

- **Step 4: Implement**

  The implementation stage, step 4, is where high capital investments will be made to deploy the implementation of smart manufacturing technologies, automation, robotics, IoT, and the connected factory of the future. The company must rethink its business model and supply chain systems in the new digitally connected ecosystem.

  This new system will deliver better customer responsiveness with digitally connected platforms. Productivity increases with the implementation of connected factory systems. The lead time is reduced due to intelligent monitoring and alert systems that trigger quick responses to solve problems. Better process quality with digital and visual monitoring systems. Increased machine utilisation with factory monitoring and rapid response time. The opportunities for customer co-creation of products with shorter time to market data-driven value (end-to-end) and human-robot collaboration. All these will deliver very good returns on capital investments with increased market share.

5 Steps to Implementing the Factory of the Future

1. **Embrace automation and digitalisation.** Incorporate advanced technologies such as robotics, artificial intelligence (AI), internet of things (IoT), and cloud computing into your manufacturing processes. This will enable you to improve operational efficiency, reduce costs, and enhance quality.

2. **Implement data analytics and real-time monitoring.** Collect and analyse data from various sources within your factory to gain insights and make data-driven decisions. This will help you optimise

production processes, predict maintenance needs, and identify areas for improvement.

3. **Adopt agile manufacturing practices**. Implement agile methodologies and lean principles to increase flexibility, responsiveness, and adaptability in your manufacturing operations. This will allow you to quickly adapt to market changes, customise products, and reduce lead times.

4. **Focus on sustainability and green manufacturing**. Implement eco-friendly practices and technologies to minimise your environmental impact. This can include energy-efficient machinery, waste reduction strategies, recycling initiatives, and usage of renewable resources whenever possible.

5. **Invest in Workforce Development and Collaborative Culture**. Develop a skilled and adaptable workforce through continuous training and upskilling programs. Foster a collaborative culture that encourages innovation, teamwork, and open communication. This will enable your employees to embrace new technologies and lead the digital transformation in your factory.

By following these steps, you can transform your manufacturing facilities into a factory of the future, equipped with advanced technologies, data-driven decision-making, and a sustainable and agile approach to production.

There must be a clear vision and objectivity connected in the new business model that will be driven by higher levels of customer responsiveness and organisational agility to compete globally. Implementing FOF or smart manufacturing is not about spending capital investment to catch up with the trends of the Industrial Revolution to show robots in action or to look elegant. Smart manufacturing is about delivering sustainable, better business results. Smart manufacturing is about being globally competitive, growing business with faster response lead time, and having a responsive supply chain and digitally integrated with your customers and suppliers, all this being delivered at a lower cost product cost.

Transitioning to Factory of the Future

The figure below explains the process of transition from the current state of the factory to the visionary factory of the future. The company, in its current state, must be strengthened with lean manufacturing implementation to create a lean organisational culture.

The key principles of lean are to continuously improve process efficiency, predictable machines, customer responsiveness, and problem-solving; eliminate non-value-added activities and continuous cost reduction; and be agile. These are important foundational prerequisites for any company to succeed in manufacturing or business. It is the readiness prerequisite required before the company can start to utilise smart manufacturing and at the same time continuously strengthen the lean manufacturing implementation.

The organisation must also periodically complete the value stream mapping (VSM) to identify opportunities to improve, eliminate non-value-added activities, and enhance the value within the company which ultimately offers increased value for the customer.

Factory of the future = Lean manufacturing + Smart manufacturing technologies

Any organization that applies lean manufacturing dynamics will continue to thrive by sustaining strong profitability, growth, and innovation even though unpredictability is the norm and change is constant. Lean manufacturing dynamics with strong customer focus and customer responsiveness are the key to business survival in any industry.

When a company has already developed a strong customer-focused lean manufacturing culture, implementing smart manufacturing or FOF becomes a logical next step to increase the value chain and grow the company to higher levels of customer responsiveness.

Figure 20: Transitioning to Factory of the Future (Lean and Smart Manufacturing)

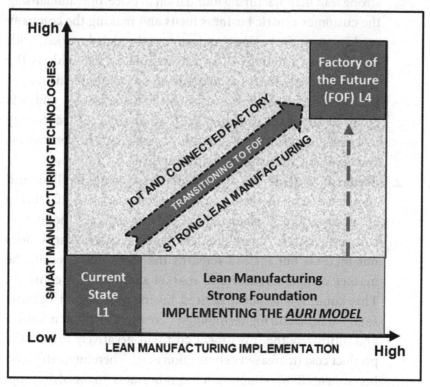

Source: Azlan Nithia, 2023

Smart Manufacturing and Market Competitiveness

The figure below explains the three types of smart manufacturing implementation scenarios and how these scenarios will have an impact on the organisation's market competitiveness, customer responsiveness, and market share.

1. **Scenario 1 (line A)**—smart manufacturing and automation implementation with IoT, robotics, and a connected factory. The increase in the level of FOF automation and smart manufacturing implementation correspondingly increases global market competitiveness. This enables the company to achieve higher customer responsiveness, greater market share, and good returns on capital investment. This is possible by having a strong lean manufacturing foundation in place, operationalising the customer-centric business focus and making the company ready to grow the business and capture the global markets with these implementations of smart manufacturing. This is the prerequisite for smart manufacturing or FOF implementation readiness and financial investments. This is line A, and it is going in the right direction with good value of technology investments and an increase in the organisation's responsiveness and competitive customer-centric agility.

2. **Scenario 2 (line B)**—smart manufacturing applied and completed. The increase in the level of smart manufacturing automation and implementation did not increase the market competitiveness. Therefore, customer responsiveness does not increase but instead remains the same. As a result, the market competitiveness and market share remain constant. This condition is not a desired business condition because smart manufacturing technologies are highly capital intensive and ultimately the investment cost will negatively impact the product cost (increased depreciation cost). Therefore, return on investment (ROI) is important in any business financial decision. In addition, scenario 2 or line B is indicative of a lack of lean manufacturing implementation as the prerequisite foundation. There should be an infusion of creativity first (reduce non-value-added activities in all the processes, eliminate unrequired activities or simplify processes) before spending money on capital expenditure. Any capital investment should increase the value of the company; the most important value is the increase in customer responsiveness (by achieving faster delivery and lower cost) and getting more business or growth globally. If we

are not growing our business, it simply means the business is not growing and creates a high risk of losing competitiveness or even closure.

Figure 21: Smart Manufacturing Investment Versus Global Market Competitiveness

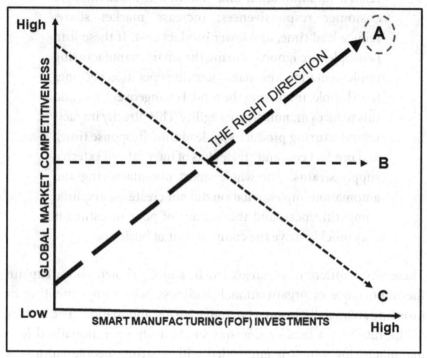

Source: Azlan Nithia, 2023

3. **Scenario 3 (line C)**—the smart manufacturing and various automation projects implemented. The increase in the level of smart manufacturing and FOF technologies that were implemented had decreased the agility or the flexibility of the internal processes and the danger of lowered global market competitiveness.

   This is a common manufacturing automation <u>TRAP</u> for many organisations embarking on the implementation of IoT, connected factories, robotics, and automation specifically. This is the situation whereby the capital is invested prior to the full lean foundation implementation. Remember creativity always comes first before any capital investment. The creativity is driven by the lean manufacturing principles. Lean requires a careful study of every capital investment related to automation and robotics; it must improve customer responsiveness, increase market share, reduce lead time, and lower product cost. If these lean principles are ignored during the smart manufacturing implementation, the manufacturing operation becomes less flexible, increasing the model changeover time, and this reduces manufacturing agility. This directly impacts manufacturing productivity, lead time (response time), and product cost and creates lots of internal and external supply strains. The whole smart manufacturing and automation implementation did not create the required competitiveness, and these kinds of poor investments may quickly drive the company out of business.

These three different scenarios (A, B, and C) shown above explain the importance of organisational readiness before implementing or investing financially in smart manufacturing automation. The factory of the future is a factory that has successfully operationalised lean manufacturing which is embedded with a strong people culture of continuous learning and improvement with efficient manufacturing processes, well-maintained equipment, and reliable processes. Anything money that an organisation invests in, especially involving automation, robotics, IoT, and capital investments, it is important to have in mind to ultimately increase customer responsiveness by intensifying flexibility, agility, faster delivery lead time, lower product cost, and better quality.

To compete successfully in the global market means to be successful in the fierce customer-centric business. It is important that the organisational readiness assessment is achieved as explained in the 'four steps to become the factory of the future (the AURI Model)'.

Achieving Factory of the Future Status—the Challenges to Achieve it!

Achieving the status of a 'factory of the future' in manufacturing requires overcoming several difficulties. Some of the common challenges include

1. **Technological integration.** One of the primary difficulties is integrating the latest technologies into the existing manufacturing processes and infrastructure. This includes implementing smart manufacturing technologies like the internet of things (IoT), big data analytics, artificial intelligence, robotics, and automation systems.

2. **Skill gap.** Transitioning to the factory of the future often requires a highly skilled workforce to operate and maintain advanced technologies. However, there is a significant skill gap in the manufacturing industry, making it challenging to find and train employees with the necessary expertise.

3. **Data management and cybersecurity.** The factory of the future relies on data collection, analysis, and decision-making based on real-time information. Establishing robust data management systems and ensuring cybersecurity to protect sensitive information from breaches and cyber threats are hurdles that manufacturers face.

4. **Capital investment.** Implementing advanced technologies comes with a significant financial cost. It requires substantial investment in upgrading equipment, integrating software systems, training employees, and managing the transition period, which might cause financial constraints for some manufacturers.

5. **Change management.** Transforming traditional manufacturing processes into the factory of the future may involve a cultural and organisational change. Resistance to change from employees and stakeholders can present challenges in terms of adopting new practices, adjusting job roles, and managing the transition effectively.

6. **Connectivity.** Establishing a seamless and reliable network infrastructure to support real-time communication and

connectivity between various systems, machines, and devices is essential for the factory of the future. However, ensuring consistent connectivity throughout the manufacturing facility can be challenging, particularly in large or remote locations.

In conclusion, addressing these difficulties requires a strategic approach, proper planning, collaboration between stakeholders, and investment in both technology and workforce development.

Conclusion

The new industrial revolution in manufacturing is focused on achieving 'smart manufacturing' and the 'factories of the future' as the prerequisite or competing globally. Achieving this requires the implementation of a 'connected factory', smart manufacturing automation, and extensive IOT technologies. The machines, robots, and processes must be connected, able to deliver real-time comprehensive analysis, and be monitored at all times to deliver world-class process performances and productivity.

To reimagine the manufacturing industry through the successful adoption of greater digital technologies in order to increase efficiencies, speed, and agility, the industry must acquire the ability to be able to rapidly transition into the smart manufacturing paradigm by adopting the latest relevant automation technologies. This will give the companies the ability to quickly respond to demanding market changes and customer responsiveness in a digitally connected marketplace. This can only be achieved by implementing higher levels of manufacturing competitiveness and a strong lean foundation.

The manufacturing industry is constantly evolving from the use of an intensive labour force to the use of automation and robotics to increase production efficiency, reduce lead time, and cut the cost of products. An efficient manufacturing organisation is a way to enhance the industry's productivity, and this requires the support of a team with a high-skilled talent pool, innovation, and continuous learning capacity. This scenario ultimately supports the well-being of the country's national economic prosperity driven by our people.

Manufacturing Industry Challenges

The manufacturing industries are constantly confronted with serious issues like rising labour costs and lacking skilled talents or employees who can learn and adopt and implement the latest digital technologies rapidly, high capital investment, a globally challenging business environment, and increasingly demanding customer requirements. These industries must constantly be capable of evolving and transforming by adopting and applying the latest efficient technologies to meet the changing and challenging customer requirements. The ability of the organisation to quickly transform itself to deliver shorter lead times, better quality products, and lower production costs will ultimately decide whether the organisation will survive or get eliminated in global competitiveness.

As manufacturing companies investigate their next decade of growth and business sustainability, they must rethink how they are currently doing business and discover the gaps. Only through implementing higher technology manufacturing methods, reinventing, innovating new systems, and running an intelligent smart operation can they stay relevant and continue to be highly competitive in the global business environment. There is no other way!

The organisations must likewise continuously focus on improving and strengthening their lean manufacturing foundations (both in people and processes) and develop a continuous learning culture so that smart manufacturing applications can generate the required customer responsiveness, organisational proficiency and people readiness to succeed in achieving the factory-of-the-future status. The organisation's relentless journey to remove non-value-added activities by applying and eliminating the 3M wastes (*muda*, *mura*, and *muri*) are basic requirements of lean to improve productivity and reduce lead time and cost of product.

People Talent Challenges

The transition to the new smart manufacturing paradigm is driven by two key pillars, namely people and process readiness, and then technology or automation. Both people and process excellence will deliver operational excellence. People readiness is a priority because only capable persons in your organisation with relevant skills and knowledge (the right people with the right skills) will succinctly adopt and implement efficient processes. What is critical is developing the required human capital by retaining the existing talents and providing them with the right skills and technical training. It is the people who will find the right technologies, implement them, and deliver the required business results to compete.

The People Talent Requirements

- Talent and skills—the ability to continuously learn, adopt, and implement new technologies rapidly.
- The people culture and readiness to adopt and manage processes in a smart manufacturing environment.
- Training needs for the new engineers and technicians to be able to maintain and improve smart manufacturing equipment and processes.

Appendix 1

Leadership in Action: Facilitating Learning Transfer for Performance Improvement

Performance Improvement

Written by Maj Dr J. Prebagaran, PhD (founder and CEO, SMC Professional Center for Learning & Development)

In every corner of the business world, the message is the same: talent is paramount. It's a message that resonates in boardrooms and manufacturing floors alike, underscoring the widespread belief in the power of human capital. Yet, despite the consensus on the importance of investing in talent development, there's a significant disconnect when it comes to actual outcomes.

The reality, as highlighted by a 2018 Association of Talent Development report, is that less than 20% of learning from training sessions is effectively applied to improve job performance. This statistic poses a critical question: What's happening to the remaining 80%? It's this question that has propelled me into a deep dive into the intricacies of learning transfer, exploring how to bridge the gap between knowledge gained and knowledge applied.

From Ancient Wisdom to Modern Insights

My exploration into the dynamics of learning transfer with inspiration from contemporary educational learning theory and the ancient wisdom encapsulated in the Tamil literature, 'Thirukural' by Thiruvalluvar. Verse 391 of this timeless text states, *'karka kasadara, karpavai katrapin, nirka atharke thage'*—which translates to, 'Learn thoroughly what needs to be learned, and then live accordingly.' This ancient wisdom strikingly parallels the contemporary challenges in learning transfer, underscoring the timeless relevance of integrating learning with action.

The issue of 'wasted' training efforts is not new. Scholars like Dr. Ina Weinbauer-Heidel have extensively documented the challenges and solutions surrounding learning transfer. Participating in a certification program based on her insights was a revelation for me, highlighting the critical gap between acquiring knowledge and applying it effectively.

Reflecting on my experience in the manufacturing sector, I've seen firsthand how lessons in waste management can apply to learning transfer. As a Manufacturing Supervisor at Mattel more than 3 decades ago, the challenges of managing physical waste provided a clear parallel to the 'waste' of untapped human potential. In a field where efficiency and waste reduction are paramount, the insights gained from managing physical waste are directly applicable to the realm of talent development.

Learning from the Manufacturing Floor

The principles of Lean Manufacturing, with its focus on eliminating the '8 Deadly Wastes,' offer valuable lessons for managing learning transfer. While physical wastes like overproduction and defects receive meticulous attention, the waste of unused talent often goes unnoticed. This oversight is particularly striking in an industry that diligently monitors even a 1% product scrap rate, yet seems to overlook the vast amounts of 'Learning Scrap'—the untapped potential of learning not applied in the workplace.

As Learning & Development professionals, we must acknowledge our role in this oversight. Perhaps we haven't been vocal enough about the critical importance of managing 'learning scrap' effectively. It's time for a change, and as with all change, it starts with ABC—Awareness Before Change.

A Call to Action for Leaders

The challenge of learning transfer is not just an educational issue; it's a leadership imperative. Effective leadership is crucial at all stages of the training process. From setting clear goals before training to supporting real-time application during sessions, and finally, ensuring the integration of new skills into daily operations post-training.

For example, consider a scenario in a manufacturing plant where a new piece of equipment is introduced. The success of this investment doesn't just depend on the technical training provided to the operators. It hinges on the leadership's ability to facilitate the transfer of this learning into the day-to-day operation of the equipment, minimizing downtime, and optimizing productivity.

A New Perspective on Talent Development

In the manufacturing world, even a 1% scrap rate draws immediate management attention. It's disturbing and disappointing, then, that in many cases, up to 80% of 'learning scrap' doesn't receive the leadership attention it desperately needs. This must change. As leaders, we must shoulder part of the responsibility for this oversight. Perhaps we haven't done enough to emphasize the significance of managing 'learning scrap.'

The pages that follow in this chapter are dedicated to offering suggestions for leaders to facilitate learning transfer more effectively. By prioritizing the management of 'learning scrap' with the same diligence applied to physical waste, leaders can unlock the full potential of their talent development investments. Let's transform the way we approach learning transfer, turning our focus from awareness to action, and in doing so, pave the way for a future where talent development is not just an investment but a tangible driver of organizational success.

To remain competitive in today's business landscape, it's crucial for leaders to champion the effective application of learning for performance enhancement. Unapplied knowledge is a waste—a luxury no forward-thinking organization can afford. Just as the manufacturing sector meticulously manages product scrap, it's time we address the overlooked issue of 'learning scrap' with equal diligence. This chapter is dedicated to empowering leaders with strategies to minimize this waste, ensuring that learning investments translate into tangible improvements.

Imagine this. Have you ever left a training session brimming with enthusiasm to implement new strategies, only to find yourself reverting to old habits weeks later? You're not alone. This scenario underscores a significant challenge in the corporate training arena: the gap between learning and its application in the workplace.

Robert Brinkerhoff, a prominent figure in learning effectiveness and evaluation, recognized this issue. Despite the substantial investments companies make in training programs, the anticipated changes often fail to materialize. Brinkerhoff's research revealed a startling fact: less than 20% of training participants effectively apply their newfound knowledge at work. This disconnect not only undermines the primary objective of learning and development effort, but also represents a poor return on investment. Individuals, organizations and nations are investing huge amount of money and other resources for the noble intention of talent development. Unfortunately, many not even aware of the huge waste due to poor application of learning.

The essence of learning and development transcends the delivery of exceptional training; it's about ensuring that these efforts yield significant results. How, then, can leaders bridge this gap, ensuring that the focus remains on the practical application of learned skills for performance improvement? It boils down to the effective management of training, encompassing pre-training design, implementation, and post-training support.

Traditionally, the responsibility of training management falls within the domain of Training Departments (or other names like Learning & Development, Talent Development, etc). While these departments' role is critical, their capacity to influence performance outcomes after completing training is often limited. This is where leadership comes into play.

Leaders have a critical role in integrating the three key phases of training. I have summarized **seven** actions leaders can take to integrate these inter-related phases to translate learning into RESULTS.

Figure 1 is an overview of the 3 Inter-related Phases of Training.

Figure 22: Leadership in Action

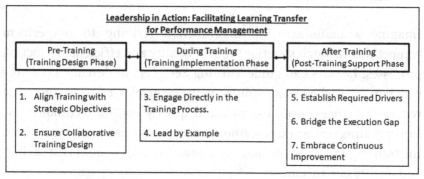

Source: Dr J. Prebagaran (2024)

Pre-Training (Design Phase)

Integrating talent development within an organization is pivotal, particularly in the manufacturing sector where the stakes for quality improvement and cost reduction are high. These elements are not just operational goals; they're the key for competitive advantage in a global market. Achieving these objectives requires a strategic approach to the pre-training design phase, ensuring that training programs are not only relevant and actionable but also directly contribute to enhancing the organization's competitive edge.

The pre-training phase is crucial for setting the direction of talent development efforts. It's where leaders have the opportunity to align training with the organization's strategic imperatives of quality improvement and cost reduction. This alignment is essential for ensuring that the training delivers on its promise of driving tangible improvements in performance and competitive advantage.

What can leaders do?

Action 1: Align Training with Strategic Objectives

Leaders must begin by explicitly communicating their expectations regarding quality improvement and cost reduction to learning designers. This involves defining clear, measurable objectives that the training program should support.

Imagine a manufacturing organization striving to outperform competitors through superior product quality and efficient production processes. Leaders can guide learning designers to develop a training program focused on lean manufacturing principles, emphasizing waste reduction, process optimization, and quality control techniques. By incorporating real-world case studies of successful lean implementations, the training can illustrate how these practices lead to both cost savings and quality enhancements.

Why this action is important?

Aligning training with strategic objectives ensures the training program is directly linked to the organization's growth strategy. It transforms training from a theoretical exercise into a practical toolkit for employees to contribute to the company's competitive advantage. By focusing on quality and cost, the training addresses the critical drivers of success in the manufacturing sector.

Action 2: Ensure Collaborative Training Design

To create training programs that truly resonate with the needs of the organization and its employees, leaders must facilitate the involvement of all relevant stakeholders in the training design process. This includes not only the learners but also those who are directly impacted by the training outcomes.

For a training program aimed at quality improvement and cost reduction, involving a cross-section of employees. Imagine the wisdom production operators can share based on their insights of thousands of hours of working on a process. These operators may not use sophisticated language. Their communication skills and confidence may be very limited. However, it will be naïve not to value the potential wisdom from their experience.

Leaders might organize workshops or focus groups to gather insights into daily operational inefficiencies, common quality issues, and potential areas for cost savings. This feedback becomes invaluable in tailoring the training content to address specific organizational needs. Imagine the value when training designers get insights, opinions, suggestions from all relevant stakeholders. The comprehensive view from multiple dimensions of inter-related activities in an organization are valuable to create impact and meaningful training program.

Collaborative design ensures the training program is rooted in the reality of the organization's operations. It enables the training to focus on practical skills and knowledge that employees can apply directly to their work, leading to immediate improvements in quality and efficiency. Moreover, this approach fosters a culture of continuous improvement, where employees feel valued and engaged in the company's success.

While the ideas of collaborative design seem logical, the challenge is executing these ideas. The training department may face challenges to get the support of various other relevant departments. Members of these department may have their own challenges and training design could be the least of their priority.

Leaders can be the catalyst to facilitate authentic collaboration of relevant stakeholders.

Leadership is critical throughout the pre-training phase. Beyond setting strategic objectives, leaders must actively promote a culture of collaboration and continuous learning. This means not only facilitating stakeholder engagement in the training design but also championing the importance of the training program across the organization.

Leaders should communicate the strategic importance of the training program, linking it to the organization's broader goals of quality improvement and cost reduction. By highlighting how the training supports these objectives, leaders can motivate employees to participate actively and apply what they learn.

The pre-training phase is more than just the initial step in the training process; it's the foundation upon which the success of talent development efforts is built. By aligning training programs with the strategic goals of quality improvement and cost reduction, and ensuring a collaborative design process, leaders can significantly enhance the impact of training. This approach not only drives immediate benefits in terms of operational efficiency and product quality but also positions the organization for sustained competitive advantage. Through strategic leadership and collaboration, training in the manufacturing sector can become a powerful tool for organizational growth and success.

By actively participating in these phases, leaders not only enhance the effectiveness of training programs but also contribute to a culture that values continuous improvement and practical application of knowledge. This approach not only reduces 'learning scrap' but also positions organizations to reap the full benefits of their investment in human capital development. Input, insights and inspiration from leaders will serve as a catalyst to turn learning into a strategic asset for performance improvement from the design phase of training.

During Training (Implementation Phase)

Integrating learning into the fabric of an organization is a multifaceted endeavour, especially when it comes to enhancing operational quality and reducing costs in the manufacturing sector. The role of leaders in this process is pivotal, not just in setting the stage through strategic training design but also in actively participating to inspire learning transfer during the training itself. The notion of Return on Your Time (ROYT) becomes a guiding principle for leaders, emphasizing the importance of investing

time in activities that yield the greatest impact on the organization's growth and competitive edge. I believe, Talent Development, particularly the facilitating the application of learning, should be among the priority of leaders.

What can leaders do to facilitate learning transfer during the implementation phase of training.

Action 3: Engage Directly in the Training Process

One of the most powerful actions a leader can take is to be physically present during training sessions. This involvement goes beyond mere oversight; it's an opportunity to connect with employees, reinforcing the value of the training and its alignment with the organization's strategic goals. I can imagine the challenges leaders may face to manage their busy schedule. With good planning, every leader can schedule his or her time. Even when they cannot, a good leader must ensure suitable representative to attend the training. If talent development is a priority, leaders shall find the time for activities with high ROYT like training.

What can leaders do during training?

i. Engaging with Learners

Imagine a leader stepping into a training session focused on Lean manufacturing—a methodology aimed at organizing and managing

209

the workspace to improve efficiency and reduce waste. The leader shares why this particular training is crucial, not in abstract terms but by linking it to the organization's immediate goals of quality improvement and cost reduction. They discuss specific organizational challenges that Lean Manufacturing can address, such as reducing production errors, minimizing downtime, and streamlining processes to lower costs.

During this engagement sessions, leaders shall express their appreciation for the insights and inputs from the learners and other relevant stakeholders. This engagement doesn't just highlight the relevance of the training; it also shows learners that their insights and contributions during the design phase were valued and incorporated. Such recognition fosters a sense of ownership and accountability among employees, motivating them to actively participate and apply what they learn.

ii. Leveraging Personal Experiences

When leaders share their own experiences with the learning concepts being taught, especially their successes and vulnerabilities, it can significantly impact learners. For example, a leader recounts how initially skeptical they were of the 5S methodology's impact on productivity and quality. Yet, after seeing the dramatic improvements in workflow organization, error reduction, and cost savings following its implementation, their perspective shifted. This narrative not only humanizes the leader but also demonstrates the tangible benefits of embracing new learning and being open to change.

Action 4: Lead by Example

Leaders inspire by example. When they actively support and participate in training programs, it sends a powerful message about the value of continuous learning and improvement.

i. Benefits of Sharing Success Stories

Let's talk more about the benefits of implementing foundational principles like 5S. A leader's sharing of how the application of 5S principles transformed their own work area or department, with specific examples, can be incredibly motivating. They might describe the before-and-after scenarios of a particular production line that was reorganized according to 5S principles. Before the intervention, misplaced tools and materials frequently caused delays and defects. After applying 5S, the area saw a significant drop-in production time and quality issues, directly contributing to cost savings and enhanced product quality.

This kind of storytelling not only illustrates the practical benefits of the training but also shows that the leader is not just preaching but practicing these principles. It underscores the idea that everyone in the organization, regardless of their position, is a learner and contributor to its success.

ii. Encouraging Open Dialogue

During training sessions, leaders should foster an environment where questions and discussions are welcomed. They can encourage learners to share their concerns about implementing 5S in their work areas, including perceived challenges and obstacles. This openness not only aids in addressing potential issues before they arise but also demonstrates the leader's commitment to supporting employees in the application of new knowledge.

Furthermore, by engaging in these discussions, leaders can gain insights into the current state of operations, identify areas for improvement, and discover opportunities for further training and development. This two-way dialogue contributes to the leader's own growth, reminding us that learning is a continuous journey for everyone in the organization.

The Impact of Leader Involvement

The active participation of leaders in the training process can dramatically enhance the impact of talent development efforts. It not only boosts the morale and motivation of employees but also ensures that the training is directly linked to achieving strategic objectives. Leaders become the bridge between learning and practical application, facilitating a culture where continuous improvement is valued and pursued.

Leaders play a crucial role in inspiring learning transfer during training. By being present, engaging with learners, sharing their experiences, and modelling the principles taught, they can significantly influence the success of training programs. This involvement demonstrates a commitment to quality improvement and cost reduction, reinforcing the message that talent development is a strategic priority.

Investing time in these activities yields a high ROYT, enhancing the organization's competitive edge and ensuring its survival in a challenging market. Through their actions, leaders not only inspire individuals to embrace and apply new learning but also foster an environment where continuous improvement and excellence are the norm. This approach to leadership and talent development is essential for any organization aiming to thrive in the fast-paced world of manufacturing.

After Training (Post Training Support)

In the realm of talent development, particularly within the high-stakes environment of manufacturing, the journey from theoretical learning to practical application is like nurturing a seed into a fruitful tree. My passion for understanding and applying this process has deep roots, beginning in the 1980s as a young undergraduate immersed in the world of Learning & Development. It was then that I was first introduced to the Kirkpatrick Model, an encounter that would shape my approach to training and evaluation for decades to come.

Donald Kirkpatrick's framework for evaluating training effectiveness became a guide for me, guiding my efforts to ensure that learning translated into tangible improvements in the military and later, in the high-precision environment of shipbuilding for the Royal Malaysian Navy. This model, particularly its evolution into the New World Kirkpatrick Model, emphasized not just the importance of learning but the critical need for learning transfer—the process of applying training to achieve significant operational outcomes. I recently had the privilege to attend the New World Kirkpatrick Certification program. An amazing learning experience to apply Kirkpatrick Model to facilitate learning transfer.

Let's explore what can leaders do during the post training phase.

Action 5: Establish Required Drivers

The concept of 'required drivers'—elements that reinforce, monitor, encourage, and reward performance of critical behaviours learned in training—resonated deeply with me. As a leader in various capacities, I recognized that the journey of learning doesn't end with training; it begins anew in the workplace. Implementing these drivers meant developing systems that didn't just expect application but actively supported it.

In the context of manufacturing, where the goals of quality improvement and cost reduction are paramount, the required drivers take on a tangible form. For example, after training teams in Lean manufacturing principles, I would encourage establishing metrics for waste reduction and process efficiency. Regular review meetings and recognition programs for innovative cost-saving measures shall became part of ensuring the seed of training found fertile soil.

Action 6: Bridge the Execution Gap

Creating these systems is only half the battle. The execution gap— where well-intended plans fail to materialize into action—is a common pitfall. My experiences in both military and civilian sectors taught me the value of commitment to these processes. In the military, the rigor and discipline in applying what was learned in training to operational strategies were critical. In the shipbuilding project, it was about adapting these principles to a civilian context, ensuring that the systems and processes developed were not just implemented but lived by everyone involved.

Have you experienced organization with world class certifications like ISO 9001: 2015 QMS who failed to execute the systems. I have seen many organizations that 'decorate the wall' with certifications yet fail to implement the system. I am sure leaders will know that systems will only create results with execution.

Action 7: Embrace Continuous Improvement

Adaptability and continuous improvement have been my guiding principles, learned from the ground up. The Centre for Army Lessons Learned (CALL) in the U.S. was a pivotal experience for me, showcasing the power of After-Action Reviews (AARs). I was introduced to CALL while attending a military training in USA in early 1990s. This practice of learning from every action, successful or not, and adapting accordingly was transformative. It taught me that systems and processes require constant evaluation and refinement to remain effective in dynamic environments like manufacturing.

The essence of AARs is cultivating a culture that prioritizes learning and innovation, understanding that risk-taking is an integral part of growth and improvement. Drawing from my military background, where risks are calculated and necessary for advancement, I applied the same mindset to the civilian world. Encouraging teams to experiment within the framework of risk management, to not fear failure but to learn from it, fosters an environment where innovation thrives.

Reflecting on my journey from a student of the Kirkpatrick Model to a practitioner in diverse and challenging environments has reinforced my belief in the power of effective training evaluation and application. The actions outlined here, drawn from personal experiences and the foundational principles of the New World Kirkpatrick Model, offer a roadmap for leaders in the manufacturing sector and beyond. By focusing on establishing required drivers, bridging the execution gap, and fostering a culture of continuous improvement and adaptability, leaders can ensure that the seeds of training grow into fruitful outcomes, enhancing quality, reducing costs, and maintaining a competitive edge in the global market.

The path from theoretical learning to practical application is complex, filled with challenges and opportunities for growth. Yet, with the right approach, informed by proven models and enriched with personal experience, leaders can navigate this journey successfully, transforming training from a mere activity into a catalyst for sustained organizational excellence.

As we conclude this chapter, 'Leadership in Action: Facilitating Learning Transfer for Competitiveness,' I find myself reflecting on the journey we've embarked upon together. The seven actions outlined across the pre-training, during training, and post-training phases are born from my personal experiences and observations in the field. These are not prescriptive mandates but rather insights gleaned from a career dedicated to enhancing organizational learning and performance. My intention is not to assert these actions as the only path to success but to share them as a framework that has, in various contexts, proven effective in facilitating learning transfer, enhancing competitiveness, and reducing the wasteful overlook of talent development.

S/No	Action	Description
Pre-Training (Training Design)		
1.	Align Training with Strategic Objectives	The foundational step where leaders ensure that training efforts are not just activities but strategic investments aligned with organizational goals.
2.	Ensure a Collaborative Design Phase	By involving a spectrum of stakeholders in the training design, leaders facilitate programs that are relevant, targeted, and embraced by those who stand to benefit the most.

During Training (Implementation Phase)		
3.	Engage Directly in the Training Process	Leader's presence and active participation in training sessions reinforce the value of the learning being imparted and model the behaviour leaders hope to see in others.
4.	Lead by Example	Demonstrating the practical application of training concepts in leaders own work sends a powerful message about the importance of continuous learning and application.
After Training (Post-Training Support Phase)		
5.	Establish Required Drivers	Creating structures and systems that support, reinforce, and reward the application of new skills ensures that learning translates into behaviour and organizational practices.
6.	Bridge the Execution Gap	The best plans for change and improvement can falter without diligent execution and follow-through. Leader's role is to ensure that the path from learning to application is clear, supported, and actively managed.
7.	Embrace Continuous Improvement	In a world where change is only constant, leaders shall facilitate and inspire lifelong learning and development be dynamic, responsive, and ever evolving.

I share these actions with humility, fully aware that learning transfer strategies are not a one-size-fits-all endeavor. Every leader brings their unique perspective, experience, and style to the table. What works in one context may need adaptation in another. Yet, it is my hope that within these actions, you find inspiration, ideas, or even validation of your own practices that can be applied or adapted to your organizational context.

The underlying aim of this chapter is to spark a conversation about the value of learning transfer—not as an abstract concept but as a tangible, strategic asset that can significantly impact an organization's competitiveness and success. Just as we meticulously manage product scrap to ensure efficiency and sustainability, so too must we approach 'learning scrap' with the same vigor and intentionality. The potential of our people—their skills, insights, and capabilities—is an invaluable resource. To overlook or underutilize this talent is a waste we cannot afford, especially in today's hyper-competitive environment.

Let these seven actions serve as a call to action for all leaders. Whether you find resonance with one, several, or all of the ideas presented, the essential message is to recognize the critical role you play in transforming learning into a competitive advantage. By actively engaging in each phase of the training process, from design through to execution and beyond, we not only enhance our organizational capabilities but also foster a culture where learning, growth, and innovation are valued and pursued by all.

In sharing these insights, my goal is not to prescribe a universal solution but to contribute to a broader dialogue about the importance of effective learning transfer in achieving organizational excellence. As we each continue on our leadership journey, may we remain open to learning— not just for our teams but for ourselves—and committed to nurturing the seeds of potential that lie within our organizations. Together, let's turn learning into action, and action into RESULTS.

References

Beer, M., Finnström, M., and Schrader, D. (2016) 'Why leadership training fails—and what to do about it', *Harvard Business Review*.

Brinkerhoff, R. O. (2006) 'Telling training's story: Evaluation made simple, credible, and effective'. Berrett-Koehler Publishers.

Diaz, S. M. (2000) 'Tirukkural with English translation and explanation'. Ramanandha Adigalar Foundation.

Kirkpatrick, J. D., and Kirkpatrick, W. K. (2019) 'Kirkpatrick's four levels of training evaluation'. Association for Talent Development.

Weber, E. (2018) 'The missing link in learning: Transfer'. Association for Talent Development.

Weibauer-Heidel, I. (2018) 'What makes training really work'. Tredition GmbH.

Appendix 2

Problem-Solving Case Study: A Case of the Porous Castings

As told by Prof Dr Shrinivas Gondhalekar (aka Dr G)

The customer had finally agreed to the sampling procedure, but the tension persisted. Actually, the president of an auto ancillary manufacturer was never free from the fear of customer complaint. Just one complaint was enough. It could have serious ramifications not just for the company but also for the parent conglomerate. One fatal car crash, attributable to crankshaft failure, reported in the press would suffice to seal their fate. As auto ancillary manufacturer was the sole supplier of crankshafts, they could not even pass the buck to anyone else! The president knew exactly how Damocles must have felt about that sword.

The issue was porosity in the castings. Crankshafts were manufactured by pouring molten, ductile-grade iron into sand moulds. The resultant castings were machined and sent to the automobile company. The president knew that some of the castings displayed porosity deep inside. A porous crankshaft was weak and liable to break when the engine was running. Although the incidence of porosity was low, only one weak crankshaft was enough to cause an accident. In the emerging world of zero defects, liability for product failure was assuming alarming proportions. Claims ran into millions of dollars these days. And even if the company survived the financial blow, it could never hope to recover lost ground with the customer.

They had tried everything possible to correct the situation. They had even hired a consultant, who was an international foundry expert. He had suggested that they invest $5 million in changing the sand system. His calculations, based on overall operations of the entire foundry, revealed that they were using only 4kg of sand for every kilogram of metal cast, whereas for this particular casting technology, 6kg of sand were recommended. Insufficient sand in the mould allowed the liquid metal to cool and solidity too fast. It was well known in foundry technology that porosity was caused by differential rates of cooling within the casting. Metal has a propensity to shrink while solidifying, creating pockets of vacuum. Normally, molten metal from adjacent areas flows in to fill these gaps, resulting in a solid block of metal, which enjoys excellent strength. If, however, metal in the adjacent areas has already solidified, the gap created by the shrinking remains empty, thus creating what metallurgists call porosity.

There was another way porosity could occur: when molten metal is poured into the mould, it displaces the air in the mould. Mould designers provide paths, called vents, for the air to escape; but if these are inadequate, some air remains trapped inside, making the casting porous. Porous metal is spongy, lacks adequate strength, and permits oil and gases from one side to leak to the other. Good crankshafts are solid. They have no porosity.

The sand-system solution was far too expensive and would take nearly a year to implement as it involved rebuilding almost half the factory. Sand occupies a large volume of space and has to be transported to the mould-making area; once the castings solidify, it has to be broken away, cooled, and recycled. If you visit a foundry, you will find that the largest equipment visible usually pertains to sand silos and the sand-handling system.

But the company had no alternative. Consultations with other foundry experts had only confirmed that more sand was the solution. After many discussions with the chief executive officer of the conglomerate, it had been decided that the investment would be funded partly by the parent conglomerate group and partly by external borrowings. Inquiries had been floated, and negotiations were in progress to finalise the order.

Meanwhile, the quality control manager had come up with a countermeasure that would minimise the risk of delivering porous crankshafts to the customer. It consisted of determining which crankshaft had porosity by submitting each and every piece to an X-ray test. The technique was effective but also prohibitively expensive. The cost of castings had shot up by 30%, and the product was now making a negative contribution.

The president recalled, with a shudder, his attempt at negotiating prices upwards with the customer. The director of the auto company had been incredulous, asking him if he was aware that all vendors were only reducing prices in order to stay in business. In a suicidal moment, the president had threatened to stop making the castings if the price hike was not granted; but he was checkmated when the customer shrugged and said they would then withdraw all other products that were being sourced from the conglomerate. He had been coldly informed that a competing foundry had offered to set up dedicated operations in Thailand or India—an option that could always be considered. Beating a hasty retreat, he had managed to wrangle a small concession by touting the massive investment planned in order to improve quality. Instead of the X-ray testing, the customer had agreed to an inspection procedure, which would minimise the risk of faulty crankshafts being delivered. A sample of five crankshafts from the one hundred pieces that constituted a batch called a melt would be cut and examined in a destructive test. If none of them showed porosity, the entire batch could be dispatched to the customer. If one, two, or three of the samples showed porosity, then all the hundred would be X-rayed. The defective ones would be discarded, and only the good ones would be sent. However, if more than three samples showed porosity, the entire batch of one hundred would be discarded. This reduced the costs of doing X-rays somewhat, even though it left a finite risk. The customer had agreed with greatest reluctance.

The president knew they could not rest easy on the concession they had received. Unless this problem was rooted out, their business was in jeopardy. He knew had had to move fast.

He reached his office in a very disturbed state of mind that day; and when he saw that the quality control manager was waiting for him, he feared the worst. She was a large woman in her midforties, hawkish on the job, and unfamiliar with the art of diplomacy. The president hated her, but he also knew that he was still in his seat because she was preventing mishaps by sticking to her guns. Fearing that she had bad news, he waved her into his office and barked, 'All right, let me have it.'

She seemed faintly perplexed at this but spoke up nonetheless, 'I had been to a quality management seminar in Singapore yesterday. There was an expert from India who insisted that finding the root cause of quality-related problems did not need technical knowledge of the industry. It only needed application of hard deductive logic.'

'Okay.' The president was breathing again. Apparently, there had been no disaster. 'Well, get on with it. I don't have all day,' he said in the irascible manner he reserved for her.

'Well, I thought his approach was quite interesting. He demonstrated it with the help of case studies, where he had applied the method himself. He has done some extraordinary work in many countries across Asia. He has trained in Japan and has a great reverence for everything Japanese. I thought we could invite him to our company to attack the crankshaft problem. I checked out his credentials. He is brilliant and is quite well known in India, but he is also very expensive. I talked to him, explained our problem. He says we will recover the cost of his fees within two weeks.'

'If you think there is a reasonable chance of success, I am ready to risk it. You decide.'

The president was willing to explore any option that would save him precious time and money. The quality control manager was in her office, on the phone to India, before he could change his mind.

Dr G was scheduled to visit the auto ancillary company in early March. He thought this would be an interesting case for his ardent student, Payal. He asked her if she could join him on the visit. She was delighted at the opportunity, and they travelled together from Mumbai. On the plane, Payal was bubbling with enthusiasm. 'I hope you will be patient with me, sir,' she said. 'I am not a metallurgist, and neither am I an expert in techniques of problem solving. I shall be grateful, sir, if you would take me through the process step by step and explain the rationale at each stage.'

'Oh, I am not sure if it is a formal technique at all. I just use common sense and deductive logic. Maybe you can identify the steps since you will get a chance to see it as an outsider. I just do it on *automatic*.' She mulled over that for some time and then made notes in her little diary.

The next day, the quality control manager welcomed them warmly. She had taken an instant liking to this man in Singapore, and something about his distracted air had told her he was *the one*. 'It was like Morpheus finally meeting Neo,' she was to tell her colleagues later on.

After exchanging pleasantries with the problem-solving team, Dr G asked to be taken on a tour of the plant from the raw material receipt area to the finished product storage. 'It helps me to gain a perspective of the problem,' he told Payal. She observed that he brushed aside explanations of operating parameters but walked back and forth along the direction of the process flow, sometimes pausing to contemplate a machine in operation as if it were the *Mona Lisa*. The quality control manager watched his every move. When he had seen enough, he asked for a table to be set up near the shop floor. 'Now, please bring six to ten samples of your defective castings. Make sure they are recent samples, ideally, something produced today.'

Soon, six samples of porous castings, serially numbered, were placed on the table. Dr G picked up the first sample; and holding it up as if it were a newly acquired prize, he addressed the group. 'Look at this crankshaft. Does it have porosity?'

'Yes, it does,' they said.

'What other defect does it have?'

They stared at him blankly. Clicking his tongue impatiently, he said, 'Please look at the list of possible defects and find out which other defect is present in this very shaft.'

They could not make head or tail of it. The quality control manager was the first to comprehend. Pointing to a large board that displayed the quality parameters of the crankshaft in question, she asked, 'Are you referring to this? We check for hardness, sometimes the microstructure, perlite content, ferrite content, and the shape and spacing of the nodules. We check for blowholes, porosity or shrinkage, surface cracks, and cold shut.'

'What's a cold shut?' asked a bright voice. It was Payal. To her surprise, Dr G waved her query aside with a brusque 'Oh, never mind, I don't care what a cold shut is as long it is one of the defects.' Handing the crankshaft to the quality control manager, he said, 'Hold it in your hands. Now examine it, test it, and look up your test record. Do what you like, but tell me what other defect from your list is present in this particular sample.'

The whole group bent over the sample and started inspecting it. They declared that there was no cold shut, no surface crack, and no blowhole was visible; but they would need more time for chemical analysis of the metal.

'Let's start with the information that is readily available. If it suffices, well and good. If it does not, we will ask for more data.'

'Work with whatever data is available. Don't let incomplete data hold up the proceedings. You can always seek additional data as you go along,' Payal wrote in her small black diary. These notes would prove invaluable later on.

Other than porosity, there was nothing wrong with sample number 1. But sample number 2 had low hardness while sample 3 had a sand inclusion on the surface. The remaining samples had no defect, except porosity. Dr G was satisfied with the data. He did not need more samples.

'Porosity is not correlated with any other defect,' he whispered to his new associate. Down it went into her diary.

Picking up sample number 1 again, he asked, 'Where exactly is the porosity present in this sample? Also, please identify areas where porosity is completely absent.'

The sample was cut. Porosity was found at the junction of the flange and the shaft, in the area known as pin 1. Payal sketched the crankshaft in her diary. It is shown below in figure 3.1.

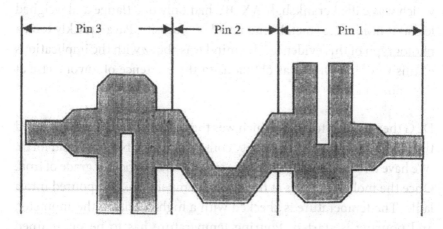

Fig 3.1 Crankshaft AX

Source: Dr Gondhalekar, 2019

Pins 2 and 3 were found to be free of porosity. The same pattern was evident in all the samples.

'Wow, the porosity appears exactly at the same spot!' Payal exclaimed with childlike enthusiasm. But Dr G was unmoved. He was like a demon possessed once the problem-solving got underway. Later, he explained to Payal that diagnosing the root cause with deductive logic required a still, meditative mind. It required a deep understanding of nature, he said, adding that the beautiful thing about nature was that it did not bend to the wishes of man. Nature's laws were absolute and immutable. Humans had to understand nature; humans had to figure out how the metal would flow and solidify.

Dr G asked for the table to be cleared. Fishing out a marker pen from his pocket, he divided the table into two halves. He requested the group to bring a sample of each product made in the foundry that had porosity as a frequently occurring defect and a sample of each product that did not have porosity as a defect. Samples were brought and placed on each side of the table. There, in front of them, was some curious evidence: that of crankshaft BY. It was a bit smaller than the one under study, which was called crankshaft AX. BY had only one flange and weighed less as it needed less metal. And it had no porosity. Payal quickly took a photograph of this evidence. Her mind was abuzz with the implications of this find, but Dr G was oblivious to the existence of anyone else at that moment.

Dr G then wanted to know which was the machine or die that produced the defective castings. The quality control manager began to hold forth, 'We have two furnaces. A furnace is specific to a particular grade of iron. Once the molten metal is at the desired temperature, it is poured into a ladle. The temperature is checked with a high-precision thermometer, and pouring is started. Pouring temperature has to be maintained between 1,400 and 1,440 degrees Celsius. One sure way of creating porosity is to pour at excessively low or excessively high temperatures. A temperature of 1,300 degrees, for instance, is sure to give porosity.

'The time between melting and pouring needs to be kept low. If molten metal is held too long, it oxidises in contact with oxygen from the air and becomes brittle. Pouring also has to be done at a fair clip. Observations have shown that pouring temperature drops from 1,430 to 1,350 degrees Celsius between the first and last moulds. I have been pressing for new better-insulated moulds, but the management is always strapped for funds.'

Dr G began to shuffle his feet, but the lady was on a roll now. It was not often that she met a man who commanded her respect, and she had to assist him. She had to tell him everything she knew.

'Pouring conditions are equally important. The rule for ideal pouring is pour quickly and uniformly with no interruptions. In answer to your query, young lady, cold shut occurs when pouring is interrupted. The metal poured earlier solidifies before the rest of the metal is poured in. At the 'junction', a discontinuity is created and presto! You have a cold shut.'

Dr G could not take it anymore. With a pained expression, he put the palms of his hands together and begged the lady to stop. He did not want to know anything more, he said. The astonished manager explained that she was only helping him by listing out the possible causes of porosity, but he was adamant.

'Information overload is as bad as or even worse than information scarcity,' he said to the group. Payal whistled softly and noted it down in a flash. 'It interrupts your logical thought processes and tempts you to stray away into specious arguments. It is the spam in your inbox, best deleted. When you have insufficient information, you will make the attempt to find what you need. If you have excess, however, you feel you know it all and you will not search. Then, no surprise, you will not find,' elaborated Dr G in his laconic style.

Dr G often remarked that, as an expert diagnostician, he brought to his client company the one thing that everybody in the company lacked: ignorance. He was quick to qualify that statement, however. It was ignorance born of wisdom, he said, as opposed to the ignorance of an illiterate fool. The problem in most companies was that too many people knew too much. They were unable to see what was right under their noses because they had too much dust in their eyes. 'I start with a clean slate every morning,' he declared. 'Pranayamas take all the stale air and previous impressions out of my system.'

The quality control manager was put off by the diagnostician's unwillingness to accept the information that she was presenting. Her image of the great diagnostician was developing a crack. But she decided to try out Dr G's way. 'Okay, I will provide only the information that you specifically ask for. What else you do you need?' she ended a bit timidly, sounding obviously hurt.

'How many dies do you have to make sand moulds of this product?' he asked, oblivious to the inflexions in her voice.

'Only one, and if you don't mind please, here we call it a pattern, not die. It is in two parts, male and female we call it,' interjected the pattern-making executive.

'How many ladles do you pour from each melt of the furnace?'

'Usually five ladles.'

'So each ladle is poured into sand moulds, and each of the sand moulds is made by the same pattern on the one and only one line that you have. Is that right?' Dr G asked.

'That is correct.'

'How do you make the mould?'

'Sand is poured into a box. The female half of the pattern makes an impression or cavity corresponding to one side of the product. Then the male half of the mould makes a cavity corresponding to the other side of the product. The two half-cavities are joined to give a complete cavity corresponding to the product. The metal is poured into the cavity. Does that suffice? We can go and see the patterns in the pattern maintenance section, if you wish.'

'Later, later. Now, only questions. After the mould has cooled, it is broken and the product is separated, right? Does it pass through only one line?'

'That is right. Only one line—I told you before. We have only one line!' Most people found Dr G's repetitive questions annoying in the extreme.

'Good! Now we are moving fast. Can you please find out if pouring from any particular ladle corresponds to more porosity defect and vice versa?'

'It will take some time. Shall we take a lunch break now?' Time had literally flown by, and they had not realised it.

'Okay, let us break for lunch, but please find out how the porosity on the AX shaft has varied with time. Look at your past records,' said Dr G.

Two hours later, they had the information. A widely held opinion was that the first ladle from each melt gave the lowest porosity in the casting because the furnace was the hottest and the sand was the coolest. Dr G refused to accept the opinion. At his insistence, data from the past few days was examined. It did not bear out the opinion. One team member went on to collect more historical data from the computerised archives. It showed that though porosity was widely prevalent, there was a period of about two months, during which fifty-four consecutive melts had shown no porosity. That was two years earlier, and one of the team members said that he distinctly remembered that it was the time when the new furnace was installed. He said that it was logical because the new furnace must have given good results, with good mixing of the additives and at the right temperature. Moreover, the new ladles that came with the furnace would have also ensured good pouring; both factors could contribute to porosity.

Payal noted another learning point in her diary: 'Get data firsthand. Look at the problem in detail before going into causes.' Dr G had put up the information on a board as it had come in. He now paused and stared at it for a long time without saying a word. The team members were busy with discussion amongst themselves, to which he was oblivious. Abruptly, emerging from his reveries, he called the group to order. 'Ladies and gentlemen, please look at the board and tell me, what do you see?' he said rather theatrically. 'Look at the data and tell me, what do you see?' he repeated.

Several comments were offered. Payal observed a curious phenomenon. Every time a cause was offered, Dr G would shake his head, refusing to accept it. He would repeat his favorite sentence, 'Look at the data. What do you see?' This went on for some time. Slowly, with an air of resignation, he turned to Payal and asked in a low voice, 'Can you see the most remarkable feature of the problem, Payal?'

Payal was pleased to be included into the discussion. 'I have remarked on it in my notebook as an interesting point, sir, but I am not sure what it implies. I have noted that the defect always occurs only in pin 1 and not in the other two pins. I also thought that the fact that it does not happen in pin 3 may be important because it looks similar to pin 1.'

'Very good. Also that it happens in AX but not BY.'

'But what does it imply, sir?'

'It means we have to look at what is different between pin 1 and pin 3 and what is different between AX and BY. Our root cause lies somewhere in the differences. Let us go and look at the pattern that makes the mould. We must examine the part of the pattern corresponding to pin 1 and pin 3 in detail and find out the differences. We must look at differences in the patterns of AX and BY too.'

The team trudged to the pattern maintenance shop and placed the male and female patterns side by side. They began to list out the differences:

1. Pin 1 was closer to the riser while pin 3 was farther away from it.

2. Pin 1 had metal entry from a thin shaft while pin 3 was at the last point of the metal flow into the thinnest shaft.
3. AX was bigger in size than BX. It would consume more metal.
4. The riser of AX was larger than that of BY.
5. The whole pattern of BY was at a lower lever from the pouring cup's position compared to AX.

As the teams were listing the differences, there was a commotion outside. One of the supervisors had come running with some startling information. He had detected porosity in crankshaft BY also! The team watched with bated breath, curious to see how Dr G would react.

For a moment, there was complete silence as Dr G absorbed the information. Then he calmly pronounced, 'Show me. Don't tell me. Show me!'

The supervisor walked out and reappeared with four samples of BY shafts—all with clearly visible porosity. Dr G looked carefully at them and then thumped the table in irritation. 'Why can't you, guys, give me correct data right in the beginning? Look, if you keep giving wrong data, nobody on earth can solve the problem. Henceforth, be careful. If you are not sure, say, 'NOT SURE!' If you don't know, say, 'DON'T KNOW'!' The entire group was startled by this sudden outburst.

'Anyway', he continued, calming down as rapidly as he had heated up, 'it makes one significant change in our approach from this point onwards. We now start looking at similarities between AX and BY. Ignore the differences! Let's get on with the job.' They continued examining similarities and differences and showing them to Dr G. After what seemed an interminable length of time, he straightened up and in a low resigned voice said there was only one really significant difference. 'Can you see it?' he asked all around.

The pattern-making expert was hardly listening. 'May I make a suggestion?' he asked. 'If we put a riser near the point of porosity, we can solve the problem because it will supply metal nearer to the point of demand during the shrinkage.'

To explain better, he brought a casting with its pouring cup, runner, and riser intact. It looked like figure 3.2.

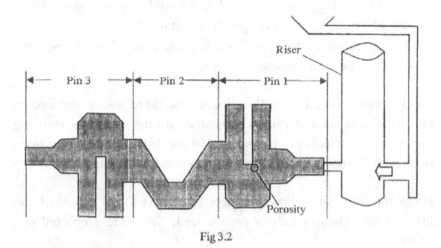

Fig 3.2

Source: Dr Gondhalekar, 2019

'Some time back', continued the pattern-making expert, 'we had faced the problem of porosity in both pin 1 and pin 3 in similar locations. I had raised the height of the riser and its volume, and the porosity in pin 3 had disappeared. I would have raised the volume of the riser even farther, but there is no space to accommodate it on the pattern. But now I have an idea. If we put another riser near the point of demand in pin 1, the porosity will disappear. I can do it by tonight.'

Dr G stopped him immediately, emphasising that no action would be allowed until completion of the diagnosis. He pointed out that the team needed to focus on one point, which was the only point of significance.

'The only significant difference between pin 1 and pin 3 lies in the distance from the riser. Pin 3 is farther while pin 1 is nearer to the riser. Their positions mean that the cooling would start from pin 3, and while it shrinks, it would draw metal from the pin 1 area, which in turn could draw from the riser. But as pin 1 solidifies, it would be unable to draw the metal it needs to make up for the inevitable shrinkage during solidification. Data shows that pin 1 is unable to draw metal from the riser.'

Dr G went on to elucidate, 'The whole question hinges on only one point: why does the riser succeed in supplying metal to pin 3 but somehow fails to supply to pin 1? The only difference between the two is that pin 3 demands the metal at an earlier point in time while pin 1 demands it at a later point in time. Now let us think over the question: what happens in that time interval?'

Slowly, as if waking up from a dream, the quality manager echoed the very words that were swirling in most people's minds. 'The riser, or its supply line, solidifies,' she whispered.

The diagnostic expert asked for a Vernier caliper, a measuring instrument. Placing it across the thinnest portion, he measured and found it to be 30mm in diameter. He marked the area with a chalk. Payal could not resist the temptation of clicking a close-up snapshot of that area with her digital camera. A sketch of it is shown in figure 3.3.

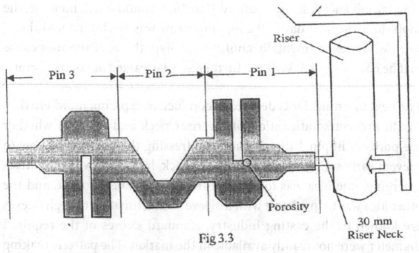

Fig 3.3

Source: Dr Gondhalekar, 2019

'Well, if thicker portions of the shaft can cool, why cannot the 30mm riser neck solidify?' inquired Dr G, adding that the root cause lay in the solidification of the riser neck happening prior to the solidification of pin 1.

One of the team members would not agree. He felt that the riser neck would be the last to solidify because it was joined to the huge mass of metal that constituted the riser. He argued that the concept of a riser was introduced in casting technology as a reservoir of metal that would remain molten till the very end and could therefore supply metal to the rest of the casting as it solidified and shrank. He said that the riser had been designed after numerous calculations and computer simulations, and it had been made even larger by the pattern-making expert to eliminate the porosity at pin 3. Early solidification of the riser or its neck was inconceivable. Moreover—and this was the clincher according to him—why did the riser neck not solidify in some cases? Why did they not get porosity on 100% of the shafts? Why was the riser neck partial to some shafts? He too was of the strongly considered opinion that an extra riser, suitably placed near the point of demand, would do the trick. He was a qualified metallurgist, and he knew what he was talking about.

Dr G refused to get drawn into the discussion. Noting that the day was now far advanced, he announced that they would reassemble on the shop floor the next day at 9 a.m. Tomorrow was D-Day, he told them. They would experimentally confirm the hypothesis of the root cause that he had proposed. Without further ado, he walked out of the factory.

The next morning, Dr G decided to conduct an experiment, where they would prevent solidification of the riser neck and confirm whether the porosity in pin 1 disappeared. Addressing the question of how to prevent early solidification of the riser neck, he found two alternatives emerging: one idea was to increase the diameter of the neck, and the other idea was to insulate it with a 'sleeve'. Unfortunately, though sleeves are known to the casting industry, standard sleeves of the required diameter were not readily available in the market. The pattern-making expert agreed to improvise a sleeve. It took him most of the day to make twenty improvised sleeves. Twenty were needed since it was decided to put one sleeve on each mould and experimentally verify for one full ladle. By 5 p.m. the sleeves were ready. Dr G then requested the team to stay back and personally observe insertion of the sleeves and entire experimental trial.

The team members distributed themselves around the pouring area. The moulds were made, the improvised sleeves were introduced, and the moulds were lined up on the casting machine, ready for the pouring. The quality manager explained to each team member what he or she had to observe. The molten metal was poured from the furnace into the ladle; the temperature was checked and found to be 1,425 degrees Celsius, which was within the specified range. Pouring was commenced. With all these preparations and so many management personnel observing and noting down his movements, the pouring operator became nervous. His pouring became non-uniform; the molten material spilt while pouring; the pouring time was high in one mould and low in another. It looked like every error in the book was occurring. When the temperature was measured after pouring the last mould, it had dropped as low as 1,270 degrees Celsius. The team was demoralised. This set of crankshafts was bound to be porous; casting technology dictated it. The quality manager courteously offered to repeat the pouring into another set of moulds.

But Dr G was unmoved. 'I don't care how the pouring was done as long as you observed how it was done. Just wait till the moulds cool, remove the castings from the sand, and check each and every one for porosity and other properties. Cut them at several places and make a thorough check. It does not matter if there is porosity. It is simply more learning,' he said and recommended they wind up for the day.

The team assembled again the next morning, except for the metallurgist, who walked in a minute late, triumphantly holding a crankshaft in his arms.

'Bad news for you, Dr G,' he beamed. 'I never believed your theory. At night, after everyone had left, I installed an extra riser near the point of demand in pin 1, with the help of my pattern-making colleagues. Well, friends, here is a crankshaft made with the extra riser, and it has no porosity! There really is no need for sleeves, as you can see!'

There was a stunned silence as every eye turned to Dr G. Surely, this was going to be the most embarrassing moment in his life. To their surprise, he calmly thanked the metallurgist, asking him to place his sample aside, and asked everyone to avoid getting distracted. 'Please continue focusing on yesterday's experimental crankshafts. Please cut them and check thoroughly for porosity.'

Meekly, the team trickled out and got down to the job of cutting and testing the samples of the previous day. Beckoning Payal to accompany him, Dr G moved with the team to the cutting machine. The pallet load of twenty castings was ready. The first casting was being loaded. 'Stop!' cried Dr G. 'Which one are you cutting?' It was number 1.

'No, no, take it down. Cut number 20 first. It had the lowest pouring temperature, below 1,300 degrees Celsius, and its pouring conditions were pretty bad. It should be the worst of the lot. Cut that first. If it turns out to be free of porosity, then we can be reasonably sure that the rest are also good. We will get an idea of the results most rapidly that way.'

The twentieth shaft was duly loaded onto the cutting machine. All the team members crowded around. Payal was anxious. The reputation of her professor depended upon what the cutting machine would reveal in the next few minutes. As the shaft parted in two, the quality manager grabbed it. She whooped with joy like a little girl and almost flung the piece into the air. 'No porosity!' she screamed.

In a voice completely devoid of emotion, Dr G said, 'Cut number 19, and go backwards till every one of them is cut.' Then he walked away to the pattern-making room and began to examine the pattern in detail. Payal followed him.

'Sir, aren't you curious about the result of the others?'

'Not at all. The results are bound to be the same. After all, I have applied stringent deductive logic to analyse the root cause. How can it be wrong?'

There he goes! thought Payal to herself. *Blowing his own trumpet again.*

Before he could say anything further, there was another commotion. A team member came rushing in, holding a cut piece, shouting excitedly, 'It has porosity! It has porosity!'

In a heart-stopping moment, Dr G had snatched the sample from his hand and was looking minutely at the porous portion. 'It cannot be,' he said with supreme confidence. 'This crankshaft could not have been made with the insulating sleeve.'

'No, no, I am referring to the crankshaft made with the extra riser!' panted the team member. 'Not the one with sleeve. I brought it from the batch that had the extra riser.'

The disgruntled metallurgist, who had installed the extra riser and who had triumphantly displayed the one good piece he had got, was cutting more pieces made from the extra riser. It was one of those that had been cut and found to be porous. Dr G's theory remained proven.

As reports of each sample being cut came in, it became clearer and clearer that the sleeve had indeed worked. Not a single one of the twenty had the slightest porosity; analysis of microstructure, hardness, and other properties revealed that those twenty were the finest-quality crankshafts ever produced from his foundry.

The metallurgist was found to have gone home due to an urgent summon from his wife. It turned out that only three out of the fifty shafts that he had cast with the extra riser had no porosity. By a lucky chance, the first one he had cut was one of the three.

Finally, everything fell in place. Dr G explained to them how they were able to draw a sample of five good pieces in the old inspection system. After the castings had cooled, when the sand moulds were broken, they passed through a sand-separating system, where the castings got jumbled up. When they came out of the system, there was no way of knowing which one was poured first. If the first few castings poured were withdrawn as samples, they perhaps tended to be free of porosity; as the temperature of the pouring dropped, the porosity increased. It was all a matter of sampling, he said, adding that this was referred to as type II error in statistical research methodology.

It was on the evening of the third day, as Dr G and Payal were returning, that she inquired, 'I admired the way you were so cool about the riser report. How were you so confident?'

'It was easy. I had a close look at the neck of the additional riser. It was small. It was bound to solidify. I was sure that the good piece was only a chance occurrence. Most people are unable to distinguish between chance causes and assignable causes. I was sure that subsequent samples would fail. And even if they did not, it would simply be a source of more data for us to understand the behavior of the metal.'

'But whatever you may say, sir, on the whole it was a brilliant performance!' Payal's face was radiant.

'Yes, it was a rather tough case. But if you hang around, Payal, this will turn out to be nothing.' With this remark, Dr G slouched in the car and began to stare out of the window at the passing scenery. He was already thinking of the next case.

Appendix 3

Deliver Breakthrough Results with Lean and Kaizen Events

Written by Dr Azlan Nithia

Kaizen is a Japanese word from the words kai and zen, which means change and better. Together it means improve to make it better or simply stated as 'continuous improvement'. Continuous improvement is extremely powerful when it is imbued into an organisational people culture. Everyone in the organisation making improvements on a daily basis will elevate the manufacturing and operational productivity constantly, where 'today is better than yesterday', 'this month's results are better than last month's result', and 'this year's results are better than last year's result', continuously improving performance.

When the continuous improvement culture is entrenched in the company's DNA, then the constant improvements will position the company as a competitive company with the capability to constantly improve and be better than the global competition.

Continuous improvement is the most powerful weapon for manufacturing excellence!

Being a lean organisation means the organisation is constantly focused on knowing the customer better, satisfying the customer, and improving the customer's responsiveness by continuously, improving the internal processes and relentless efforts to remove waste in every process in the organisation—starting from the beginning (receiving orders), making (or manufacturing) the products, delivering the orders on time in full, and collecting the payment on time as well. By doing so, the organisation must continuously reduce lead time and improve productivity (throughput or output increases), and the product cost is reduced. These improvements will increase customer delight and customer satisfaction. This results in the customer giving more orders.

Kaizen Steps

Kaizen events are normally five-day events, and they consist of five important steps, normally one step a day. This is called the full *kaizen* event. The full *kaizen* event is also called the *kaizen* breakthrough event because, in five days, the *kaizen* team implements a major breakthrough, where improvements are beyond 20%. It is common to achieve 30% improvements.

There are also smaller *kaizen* events; they are called the *kaizen* blitz. These events are for delivering quick results with the event duration being very short—between half a day up to a maximum of two days.

The Lean Thinking

- Add value to the customer.
- Create flow in the manufacturing process.
- Continuously improve and innovate.
- Reducing lead time is essential.
- Treat people as an appreciating asset.
- Establish a clear narrow focus for the *kaizen* and with clear deliverables.
- Apply breakthrough *kaizen* and deliver the results (improvements of 20% or beyond).

The Kaizen Breakthrough Focuses on

- clear objectives, deliverables, and scope for the *kaizen*.

- bias for action and delivering results rapidly
- immediate results (proven by the end of kaizen), new process functioning by day 5 and meeting the deliverables.
- using creativity to improve before spending the capital money
- Using creativity to achieve immediate process improvement
- making the physical transformation
- overcoming resistance to change
- kaizen as a team-based process
- everyone in the kaizen team being involved in the improvement
- cross-functional team members being recruited
- rapid improvement and have a tight focus on time
- necessary resources being made available immediately

Kaizen Process (Example—5-Day Kaizen)—Five Steps

Step 1 (Day 1):

Train—Lean methodology training and team exercises. Step 1 is generally about training the participants on lean principles, the 3M waste (*muda*, *mura*, and *muri*), value add (VA) and non-value add (NVA), one-piece flow, pull system, continuous improvement, managing daily improvements (MDI), standard operating procedures (SOP), JIT, *jidoka*, respect for people, lead-time reduction, and principles of kaizen breakthrough results. The kaizen objectives and deliverables are discussed, and the team leader and members are identified. Break up the members into two to three different teams with clear focus and deliverables.

Important—Most Common Failures

One common mistake that happens in many organisations when conducting or facilitating a *kaizen* event is the preconceived solution in the facilitating person's mind. He or she uses the kaizen event to get everyone to agree and validate their personally preconceived solution. These kinds of kaizen facilitators must be removed from conducting any kaizen event. These type of facilitators 'kills' the kaizen creativity of the team.

Follow the kaizen steps (SOP) strictly. Never find a shortcut to show results. Always start with a *short* training, and never ever conduct a kaizen event without a short training in the beginning. For a full kaizen, the training should be at least an hour to half a day. Training can be done by anyone in the kaizen team, not necessarily only by the facilitator.

For a kaizen blitz, the training can be between thirty minutes to an hour at least. It is common for organisations to conduct kaizen without any kind of training with the excuse that they are already been trained in the past. Never give these excuses for not doing the kaizen start-up training. Allow those members who were trained before to do the training in the kaizen. Remember <u>teaching creates the best learning</u> and reinforces their past learning.

Step 2 (Day 2)

<u>Discover</u>—analyse the current process or work (current state). The team leader and members are identified. Objectives, deliverables, and scope are clearly explained. The team is at the shop floor or the area where the kaizen is taking place. Understand all the current process, the SOPs, and how the work is being formulated. The team also identifies the waste (opportunities), inventories, reworks, and one-piece flow opportunities. The team also attempts hands-on improvements to test the improvement ideas.

The team meets to discuss the current process and all the opportunities for improvement.

The <u>first kaizen step action is to do the 5S</u>, especially complete the 2S (remove all those things that do not belong to that kaizen area and then put everything in the right place, a place for everything).

Step 3 (Day 3)

<u>Doing</u>—make the improvements. The team would have completed the 5S activities and made the workplace and the process flow visual. Determine the material replenishment system, take time, customer's

requirements, and make improvements. Implement all the improvement ideas and make corrections as required as defined in your team scope and deliverables.

The team meets to discuss the outcome of the improvements done, its results, and refinements required. Every day, at the end of the day, allocate about ten minutes for every team to give their progress update. This progress update gives every team member the progress, findings, and solutions the teams are considering.

Step 4 (Day 4)

<u>Refinement</u>—refine and improve to achieve the results. Establish the new standard operating procedure (new SOP). It is important to <u>discuss the improvements with the employees affected in the area</u> or the process to make further refinements. Install the kaizen newspaper (it could be a flip chart) to track the hourly performance of the new method and continue to refine it. Train all affected with the new change and the new SOP. Operate fully in the new system and do the hourly performance tracking and the refinements or improvements being done.

The team meets to discuss the results and compare them with the expected deliverables. Establish the visual performance tracking boards; write up the implementation of the new SOPs and new training required. Remember to do the daily end-of-the-day progress updates to all in the kaizen team.

Step 5 (Day 5)

<u>Presentation and celebration</u>—review all the new key learnings and the thirty-day action list (also called the thirty-day kaizen action list). Any outstanding actions and improvements must be completed within thirty days (or earlier). The team must also conduct the daily MDI (managing daily improvements) at the area or process that was kaizen for the next thirty days. Every day go back to the kaizen area to make some small improvements (MDI). There will be some refinements that may be required for the next several days or weeks. This will require daily

review and refinement of the new SOPs, material movements, visual performance tracking (hourly tracking of new target versus actual), and reviewing the new work methods.

Prepare the final kaizen presentation (usually in a power point slide show, PPT) and present it to the organisation's leadership team. The presentation must include the team members names, objectives, deliverables, scope, current state baseline and discoveries, improvements made, the new future state (the new results) implemented, improvements achieved (per cent of improvements and savings), and show the thirty-day kaizen action list, MDI actions, and key learnings.

Show one page that states what could possibly go wrong with the new SOP and new results and the actions to prevent those possible failures. The kaizen ends with the recognitions given to all the kaizen team members and a simple celebration of the team's success.

Improving the Kaizen Gains (MDI)

- The thirty-day kaizen action list and homework require part-time involvement by a team leader and selected team members.
- Daily *gemba* walk by relevant employees and area heads, do the MDI (making daily improvements), and make small steps of continuous improvements daily. Making small steps of daily improvements is better than trying to sustain the same performance and going backwards.
- Train area or line leaders and operators
  - Standard work, visual SOP. Implement a mistake-proofing methodology in every workstation to prevent any employee errors (apply Poka-Yoke mistake-proofing methodologies).
  - Visual real-time digital controls and performance measurement (hourly and daily tracking versus the targets). Use digital dashboards to display real-time performance results visually so that all employees in that area could see them. This allows the employees to take quick action when something is not meeting the new kaizen targets or results.

- Record countermeasures taken for any abnormality and problems solved.
  - Post a visual suggestion board (kaizen newspaper) in the work area and make sure that they are attended to daily during the daily *gemba* walk (or the MDI process).

Conducting the Kaizen Events—the Rules

- Keep an open mind to make changes and be an equal-status team member.
- Maintain a positive attitude and listen to other member's ideas.
- Never stay in a silent disagreement. Speak up and participate.
- Create a blameless environment and focus on moving forward.
- Practice mutual respect every day.
- Treat others as you like to be treated.
- Every person can have a voice, no position or rank during kaizen activity.
- There's no such thing as a dumb question. Ask if not sure.
- Quick and simple are better than slow and elegant.
- Understand the improvement process and do it immediately.
- Most importantly, strictly follow the kaizen steps SOP. Never take any shortcuts.

Kaizen Business Strategy

- People are the most important appreciable asset in the company, and respect for people is critical. The well-trained people will develop efficient processes, and continuously improve and implement well-maintained machines.
- Reduce lead time and non-value-added activities in all processes relating to new product development, manufacturing processes, and administrative processes (from end to end).
- Use value stream mapping (VSM) to reduce/eliminate non-value-added activities, reduce inventory, and improve productivity in manufacturing and business processes. The 3M waste (*muda, mura,* and *muri*) must be 'attacked', reduced, or

eliminated. Kaizen events are used to reduce and eliminate the 3M waste, the various forms of *mura* and *muri* constraints in the production processes and product flow, which are the creators of waste (*muda*).

- Continuously find and reduce or eliminate bottlenecks (constraints—mura) in the production process to constantly improve throughputs (output per hour).
- Create the culture of 5S. This must be the core culture in every function in the organisation. Make it a requirement to show the 5S score visually in every area and department.
- Digital visual performance tracking boards are critical, especially for engaging the people in that area. Tracking performance visually must include the target expected and actual results achieved and show the variance (+ve or -ve). If below target, a brief comment must be written to explain the shortfall. This must be installed and tracked every hour by all production lines and processes. There will be a large digital, visual board on the shop floor to show the daily and weekly performance summary, all the various improvement activities and problems solved in the week. This is the board that will be used during the weekly leadership meeting with the production people, and during this meeting, this is when the management and leadership team will offer any help (if required) to the team.
- Improve safety, quality, cost, and delivery of the final product to the customer (on time in full and all the time). Always discuss safety and quality first and then all other performances.
- Establish the gap towards achieving global competitiveness and performances that will enable sales and profitability to continuously grow.
- Build an organisational culture where people's involvement, engagement, mutual respect, continuous improvement, continuous learning, problem-solving, and innovative leadership are made the organisation's DNA.
- Practice problem-solving as a daily activity at every level. Problem-solving methodology must be made simple and easy

for everyone to apply. Use the A3 DD methodology. Make problem-solving a part of everyone's performance evaluation.

- Recognise small and big successes of every employee and team. Celebrate good results (small and big) and appreciate every suggestion.

Key Thoughts on Lean—Deploy the Lean Enterprise

- Manufacturing processes are all about the *flow*, the flow of parts or products from one process to another process (raw material and information input up to the completion of the finished product or the final output).
- Business processes are also about the *flow*, but in business processes, it is often the flow of knowledge.
- Problems are opportunities (like treasures newly found) for continuous improvements.
- A problem is defined as the deviation from the standard, the gap between the actual and the standard or the target.
- Sign of problems—when the management team is making decisions without the current actual *gemba* data to support them.
- Tribal knowledge is *not* enough to make smart decisions. Knowledge may be a belief, someone's feeling, or simply someone's hope. Go to the *gemba*, for truth is in the *gemba*.
- Kaizen event team members are generally from different functional backgrounds. Sometimes people onside (or different functions) will give or make honest and candid assessments of the process or business case.
- The upstream is the supplier and the downstream is the customer.
- Manufacturing is connected between the supplier (upstream) and customer (downstream), therefore, the whole system supplier—manufacturing—customer is the total value-driven supply chain (called the end-to-end value stream).
- Visual metrics are important for engaging people to achieve a common goal.

- Measure performances that truly matter for the company and make measurements visual. Every measurement must indicate the target versus the actual and state the reason for not achieving the target (if any).
- To successfully compete and sustain profitability, companies must focus on constantly eliminating waste from every manufacturing and business process.
- The foundation of a lean company is the deployment of a good 5S, visual flow, and visual performance management.
- Lean companies teach every employee in the organisation to be able to identify the 3M wastes (*muda*, *mura*, and *muri*) in everything they do.
- There are no shortcuts to manufacturing excellence. Achieving lean enterprise status requires senior leadership commitment and an organisational culture to drive it.
- Value is not about the time on the job or showing up. It is about the accomplishment and the value created by the product.
- The first step to identify opportunities and waste is to complete the value stream mapping from start to finish (end to end). Use the VSM outcome to identify all the opportunities and then use kaizen events to deep dive to materialise the savings.

Companies must become a __living business__—customers' needs will continuously evolve and shift. We must keep moving with them to ensure we stay relevant to our customers.

Actively anticipate your customer's needs by living __your business__.

Acquire an intimate 'customer knowledge' to grow your business.

When someone asks you a question—__what does your company do?__ What will your answer be? How will leaders and employees in your company answer? Examples of answers could be like 'We make PCBA', 'Car components', 'Computers', and so on. All these kinds of services are replaceable by sourcing them elsewhere.

How do we position ourselves as an <u>irreplaceable organisation</u>? If you have the kind of mentality of thinking that your company makes products or provides services, making parts and components, then you are limiting the company's ability to grow and become a customer-centric organisation.

The organisation and its community must strongly believe in and practice customer partnership, providing solutions and constantly increasing value for the customers. We depend on each other to be successful and grow both the business and market offerings. These kinds of relationships will require a deep understanding of customers and their challenges.

Applying the Lean Approach to Marketing

There is common evidence of inadequate processes and avenues for learning what the customers really need. One good example is new product designs. What are the features in the new product design that are truly important and have the ability to conjure up the customer's attention? What features create the delight factors and what are not valued by the customer? It is not about designing the product that we want but more about understanding our customers as a corollary to know their wants.

- We must have similar visions and values in tandem with those of our customers, to continually improve and grow with them and to become an integral part of our customers' success.
- The attitude and the thinking of the people in our organisation must be that the customer is always right. Regularly ask for customers' feedback and pay attention to their complaints. Take their grievances as being very serious and urgent. This is a golden opportunity to improve our product and services, to quickly solve the complaints and to show the customer how you solved their complaints with urgency and speed. Ask for more feedback and be a solution provider.

- Focus on the long-term and be growth-oriented. Look for opportunities to move the customer-centric bar towards becoming the customer-partnership. Develop an intimate understanding of the customers' businesses, challenges, and needs, and constantly commit to contributing to your customer's success. This is how a win-win strategy provides growth to both you and your customers.
- Senior leadership should commit to spending time and being involved in the customer's place and having the customer understand the company's products and service performance and opportunities to improve. This is crucial. This creates intimate customer knowledge with the sole purpose of understanding, learning and improving. This drives innovative solutions and defines the value of your company. This will differentiate your company from other traditional companies.
- Always carry out internal study (value stream mapping—VSM) to evaluate and understand the value of the activities being carried out by everyone in the company, in every function, and ask questions (see examples below) to trigger a kaizen blitz or kaizen event to eliminate any form of wasted activities and to increase value:
  - How does this activity add value for our customers?
  - Is the customer willing to pay for this activity?
  - How can we reduce inventories in every process?
  - How do we reduce lead time?
  - What are the opportunities to reduce labour, material, and overhead costs?
  - Adopt the benefits of IOT and digital technologies.
  - Does the senior leadership team understand the importance of all the above?

The lean approach to marketing is very different from the traditional approach to marketing. It starts with customer-centric questions and matching them with what you offer to your customers to meet their needs and wants to help solve their challenges. We must run a different race compared with that of our competitors, not running the same race, but by being faster and by entering areas your competitors are not ready or have the inclination to. To be a solution provider means you find a unique mix of products, services, or activities that will create superior value for the customers. An excellent example is smaller lot sizes and with higher mix of products at lower cost.

Any form of waste that exists in the manufacturing system or the administrative processes is classified as non-value-adding activities, and it deters or impedes the initiatives to achieve lead time reduction, throughput improvement, cost reduction, on-time deliveries, and quality improvements. The ultimate focus of lean is to continuously reduce the lead time to improve the flow and rapidly respond to changing customer's needs. To achieve this objective, eliminating the 3M waste is important. Equally important in the lean journey is to implement processes that are agile and flexible (for example multi-model production, modular production lines, and quick model changeovers) to meet the changing customer requirements for model variations and small lot production needs. The outcome of the lean journey will be to constantly reduce the delivery lead times and costs, which will obviously improve the organisation's competitive edge in the industry.

Customers are constantly demanding more rapid changes all the time. If you do not adapt or improve your organisation's culture to the changing customer demands, you will never meet your customers' expectations. All these requirements are the key foundations and are the important operational readiness required to implement and to be a digitally connected factory to achieve the status of the factory of the future with the ability to compete globally.

References

Abair, R. A. (1997) 'Agile manufacturing: successful implementation strategies'. *Annual International Conference Proceedings ± American Production and Inventory Control Society*, pp. 218–219.

Ahlstrom, A. P., and Westbrook, R. (1999) 'Implications of mass customization for operations management: An exploratory survey', *International Journal of Operations and Production Management*, 19(3), pp. 262–274.

Ahmad S., Schroeder, R. G., Mallick, D. N. (2010) 'The relationship among modularity, functional coordination, and mass customization', *European Journal of Innovation Management*, 13(1), pp. 46–61.

Anderson, D. M. (1997) *Agile Product Development for Mass Customization*. New York: McGraw-Hill.

Anderson, D. M. (2004) *Build to Order and Mass Customization*. Cambria, California: CIM Press.

Argyris, C. (1964) *Integrating the Individual and the Organization*. New York: John Wiley and Sons.

Atzeni, E., Luliano, L., Minetola, P., and Salmi, A. (2010) 'Redesign and cost estimation of rapid manufacturing plastic parts', *Rapid Prototyping Journal*, 16(5), pp. 308–317.

Bessant, J., Francis, D., Meredith, S., Kaplinsky, R. and Brown, S. (2001) 'Developing manufacturing agility in SMEs', *International Journal of Technology Management*, 22 (1/2/3), pp. 28–54.

Blanche, T. M., and Durrheim, K. (1999) *Research in Practice: Applied Methods for the Social Sciences*. Cape Town: UCT Press.

Boer, C. R., Pedrazzoli, P., Bettoni, A., and Sorlini, M. (2013) *Mass Customization and Sustainability*. Springer, NY.

Brandyberry, A., Rai, A., and White, G. P. (1999) 'Intermediate performance impacts of advanced manufacturing technology systems: an empirical investigation', *Decision Sciences*, 30(4), pp. 933–1020.

Brown, S. and Bessant, J. (2003) 'The manufacturing strategy—capabilities links in mass customization and agile manufacturing—an exploratory study', *International Journal of Operations and Productions Management*, 23(7), pp. 707–30.

Stump, B., Badurdeen, F. (2012) 'Integrating Lean and Other Strategies for Mass Customization Manufacturing: a case study', *Journal of Intelligent Manufacturing*, 23, pp. 109–124.

Cameron, K., and Freeman, S. (1989) 'Cultural congruence, strength and type; Relationship to effectiveness'. *Academy of Management Annual Convention*, August 1989, Washington DC.

Cavana, R. Y., Delahaye, B. L., and Sekaran, U. (2001) *Applied business research: qualitative and quantitative methods*. Australia: Wiley & Sons

Chandler, A. (1962). Strategy and Structure: Chapters in the History of the American Industrial Enterprise, Irwin, Boston, MA.

Chandler, A. (1992). 'Corporate strategy, structure and control methods in the United States during the 20th century', *Industrial and Corporate Change*, 1(2), pp. 263–84.

Christopher, M. and Towill, D.R. (2000). 'Supply chain migration from lean and functional to agile and customized', *Supply Chain Management*, 5(4), pp. 206–13.

Cigolini, R., Pero, M., Rossi, T., Sianesi, A. (2014) 'Linking supply chain configuration to supply chain performance; a discreet event simulation model', *Simulation Modelling Practice Theory*, 40, pp. 1–11.

Coates, T. D. (1998) *The Paperless Laboratory: An Integrated Environment for Data Acquisition, Analysis, Archiving and Collaboration, Trends in Research.* New York, NY: Plenum Press.

Crocitto, M. and Youssef, M. (2003) 'The Human Side of Organizational Agility', *Industrial Management & Data Systems*, 103(6), pp. 388–397.

Da Silveira, G., Borenstein, D., and Fogliatto, F. S. (2001) 'Mass Customization: Literature review and research directions', *International Journal of Production Economics*, 77, pp. 1–3.

Deal, T. and Kennedy, A. (1982) *Corporate Cultures: The Rites and Rituals of Corporate Life.* Addison-Wesley, Reading, MA.

Denison, D. R. (1990) *Corporate Culture and Organizational Effectiveness.* New York: Wiley.

Denison, D. R., Janovics, J., Young, J., and Cho, H. J. (2006) *Diagnosing organizational cultures: validating a model and method.* Working paper, International Institute for Management Development, Lausanne, Switzerland.

Denzin, N. K. (1989) 'Interpretive interactionism', *Applied Social Research Method Series.* Newbury Park, CA: Sage.

Devor, R., Mills, J. J. (1997) 'Agile Manufacturing Research: accomplishments and opportunities', *IEE Transactions*, 29(10), pp. 813–824.

Dove, R. (1995) *Presentation made at the Benchmarking for Agility Workshop.* Automation and Robotics Research Institute, Fort Worth, TX.

Dowlatshahi, S., Cao, Q. (2005) 'The impact of alignment between enterprise and information technology on business performance in an agile manufacturing environment', *Journal of Operations Management*, 20, pp. 531–550.

Duguay, C., Landry, S., and Pasin, F. (1997) 'From mass production to flexible/agile production', *International Journal of Operations & Production Management*, 17(12), pp. 1183–95.

Duray, R. (2002) 'Mass Customisation origins: Mass or custom manufacturing', *International Journal of Operations & Production Management*, 22(3), pp. 314–328.

Easterby-Smith, M., and Thorpe, R., and Lowe, A. (1991) *Management Research*. London: Sage.

Elsass, P. and Veiga, J. (1994) 'Acculturation in acquired organizations: a force field perspective', *Human Relations*, 47(4), pp. 431–53.

Feitzinger, E., and Lee, H. L. (1997) 'Mass customization at Hewlett-Packard: the power of postponement', *Harvard Business Review*, 75(1), pp. 116–121.

Felin, T., Foss, N. J., Heimeriks, K. H., and Madsen, T. L. (2012) 'Microfoundations of routines and capabilities; individuals, process and structure', *Journal of Management Studies*, 49(8), pp. 1351–1374.

Ferguson, S. M., Olewnik, A. T., and Cormier, P. (2014) 'A Review of Mass Customization across Marketing, Engineering and Distribution Domains toward Development of a Process Framework', *Research in Engineering Design*, 25(1), pp. 11–30.

Fogliato, F. S., Silviera, G. J. C. (2011) *Mass Customization; Engineering and Managing Global Operations*. Spinger, London.

Forrester, R. (1995) 'Implications of lean manufacturing for human resource strategy', *Work Study*, 44(3), pp. 20–24.

Gandhi, A., Magar, C., and Roberts, R. (2014) *How Technology Can Drive The Nest Wave of Mass Customization*. www.mckinsey.com/.../MOBT32_02-09_MassCustomization. [Accessed on 23 December 2014].

Gilmore, J. H., and Pine, B. J. (1997) 'The four faces of mass customization', *Harvard Business Review*, 75(1), pp. 91–101.

Goldhar, J. D. and Jelinek, M. (1983) 'Plan for economies of scope', *Harvard Business Review*, 61(6), pp. 141–148.

Gondhalekar, S. and Karamchandani, V. (1994) 'Robust Kaizen Systems', *The TQM Magazine*, 6(3), pp. 5–8.

Gosling, J., Purvisa, L., Naima, M. M. (2010) 'Supply chain flexibility as a determinant of supplier selection', *International Journal of Production Economics*, 128(1), pp. 11–21.

Gould, P. (1997) 'What is agility?', *Manufacturing Engineer*, 76(1), pp. 28–31.

Gunasekaran, A. and Yusuf, Y. Y. (2002) 'Agile Manufacturing; A Taxonomy of Strategic and Technological Imperatives', *International Journal of Production Research*, 40(6), pp. 1357–1385.

Hallgreen, M. and Olhager J. (2009) 'Lean and Agile Manufacturing: External and Internal Drivers and Performance Outcomes', *International Journal of Production and Operations Management*, 29(10), pp. 976–999.

Hamel, G. and Prahalad, C. (1994) *Competing for the Future*. Boston, MA: Harvard Business Press.

Hart, C. W. (1995) 'Mass Customization: Conceptual underpinning, opportunities and limits', *International Journal of Service Industry Management*, 6(2), pp. 36–45.

Hart, C. W. (1996) 'Made to Order', *Marketing Management*, 5(2), pp. 12–22.

Hayes, R. and Wheelwright, S. (1984) *Restoring Our Competitive Edge*. New York, NY: Wiley & Sons.

Hines, P., Holweg, M., and Rich, N. (2004) 'Learning to evolve: A review of contemporary Lean thinking', *International Journal of Operations and Production Management*, 24(10), pp. 994–1011.

Hofstede, G. (1995) 'Cultural constraints in management theories'. In: Wren, J. (ed.), *The Leadership Companion: Insights on Leadership through the Ages*. Free Press, New York, NY.

Huang, X., Kristal, M. M., and Schroeder, R. G. (2008) 'Linking learning and effective process implementation to mass customization capability', *Journal of Operations Management*, 26(5), pp. 714–29.

Huang, X., Kristal, M. M., and Schroeder, R. G. (2010) 'The impact of organizational structure on mass customization capability; a contingent perspective', *Production and Operations Management,* 19(5), pp. 515–530.

Huffman, C., and Kahn, B. (1998) 'Variety for sale: Mass customization or mass confusion', *Journal of Retailing,* 74, pp. 491–513.

Ireland, R. D. and Hitt, M. A. (1999) 'Achieving and maintaining strategic competitiveness in the 21st century: the role of strategic leadership', *The Academy of Management Executive,* 13(1), pp. 43–57.

Liker, J. K. (2004) *The Toyota Way: 14 Management Principles.* New York: McGraw-Hills.

Kahn, K. B. (1998) 'Benchmarking sales forecasting performance measures', *Journal of Business Forecasting, Winter 1998–1999,* pp. 19–23.

Kanter, R. M. (1992) *The Change Masters.* Routledge, London.

Katayama, H., Bennett, D. (1996) 'Lean Production in a changing competitive world: A Japanese perspective', *International Journal of Operations and Production Management,* 16(2), pp. 8–23.

Kekre, S., Srinivasan, K. (1990) 'Broader Product line: A necessity to achieve success', *Management Science,* 36, pp. 1216–1231.

Kidd, P. T. (1996) 'Agile manufacturing: a strategy for the 21st century', *IEE Colloquium* (Digest), 74, 6.

Lai, F., Zhang, M., Lee, D., Zhao, X. (2012) 'The impact of supply chain integration on mass customization capability; an extended resource-based view IEET Trans', *Engineering Management,* 59(3), pp. 443–456.

Lamming, R. (1993) *Beyond Partnership.* Prentice-Hall, Hemel Hempstead.

Lampel, J. and Mintzberg, H. (1996) 'Customizing customization', *Sloan Management Review,* 38, pp. 21–30.

Lawler, E. E. (1986) *High Involvement Management: Participative Strategies for Improving Organizational Performance*. San Francisco, CA: Jossey-Bass Inc.

Lazonick, W. (1991) *Business Organization and the Myth of the Market Economy*. Harvard University Press, Cambridge MA.

Leffakis, Z. M. and Dwyer, D. J. (2014) 'The effects of human resource systems on operational performance in mass customization manufacturing environments', *Production Planning and Control: The Management of Operations*, 25(15).

Liao, K., Ma, Z., Lee J, Y., and Ke, K. (2011) 'Achieving Mass Customization through Trust Driven Information Sharing: A Suppliers Perspective', *Management Research Review*, 34(5), pp. 541–552.

Liu, G., Shah, R., and Schroeder, R. G. (2011). 'The relationship among functional integration, mass customization and firm performance', *International Journal of Production Research*, 50(3), pp. 677–690.

Lindorff, M. (2001) 'Are they lonely at the top? Social relationship and social support among Australian managers', *Work and Stress*, 15(3), pp. 274–282.

Liu, G., Shah R., and Schroeder, R. S. (2010) 'Managing Demand and Supply Uncertainties to Achieve Mass Customization Ability', *Journal of Manufacturing Technology Management*, 21(8), pp. 990–1012.

Liu, G. and Deitz, G. D. (2011), 'Linking supply chain management with mass customization capability', International Journal of Physical Distribution & Logistic Management, 41(7), pp. 668–683.

Macaulay, J. (1996) 'Management in the agile organization'. In Montgomery, J. C. and Levine, L. O. (eds.), *The Transition to Agile Manufacturing*. ASQC Quality Press, Milwaukee, WI.

Maskell, B. H. (1991) *Performance Measurement for World Class Manufacturing*. Productivity Press, Portland, OR.

McBurney, D. H. (1994) *Research Methods*, 3rd Ed. Pacific Grove, CA: Brooks/Cole Publishing.

McCarthy, I. P. (2004) 'Special issue editorial: the what, why and how of mass customization', *Production, Planning & Control*, 15, pp. 347–351.

Medini, K., Duigou, J. L., Cunha, C. D., Bernard, A. (2015) 'Investigating mass customization and sustainability compatibilities', *International Journal of Engineering, Science and Technology*, 7(1).

Meredith, S., and Francis, D. (2000) 'Journey towards agility: the agile wheel explored', *The TQM Magazine*, 12(2), pp. 137–143.

Miles, M. B., and Huberman, A. M. (1994) *Qualitative Data Analysis: A Sourcebook of New Methods*, 2ⁿᵈ Ed., Thousand Oaks, CA: Sage.

Mostyn, B. (1995) 'The content analysis of qualitative research data: A dynamic approach'. In M. Brenner, J. Brown, and D. Canter (eds.), *The Research Interview: Uses and Approaches*. London: Academic Press. p. 63.

Narver, J. C., and Slater, S. F. (1990) 'The effect of a market orientation on business profitability', *Journal of Marketing*, 54(4), pp. 20–35.

Nithia, Azlan (2019) *Achieve Manufacturing Excellence, Lean and Smart Manufacturing*. Partridge Publishing, Singapore.

Patton, M. Q. (1990) *Qualitative Evaluation and Research Methods*. Newbury Park, CA: Sage.

Pero, M., Abdelkafi, N., Slanesi., and Blecler, T. (2010) 'A framework for the alignment of new product development and supply chain', International Journal of Supply Chain Management, 15(2), pp. 115–128.

Piller, F. T. (2001) 'The myths of mass customization', *Proceedings of the World Congress on Mass Customization and Personalization*. Hong Kong University of Science and Technology, Hong Kong (2001, October).

Piller F. T. (2007) 'Observations on the Present and Future of Mass Customization', *International Journal of Flex Manufacturing Systems,* 19, pp. 630–636.

Pine, B. J. (1993) *Mass Customization: The New Frontier in Business Competition*. Boston MA: Harvard Business School Press.

Pine, B. J. (1993) 'Mass customizing products and services', *Planning Review*, 21(4), pp. 6–14.

Pine, B .J., Peppers, D., and Rogers, M. (1995) 'Do you want to keep your customers forever?', *Harvard Business Review*, 73(2), pp. 103–14.

Pine, B. J., Victor, B., and Boynton, A. C. (1993) 'Making mass customization work', *Harvard Business Review*, 71(5), pp. 108–109.

Roethlisberger, F. J. (1977) 'Elusive Phenomena of Organizational Behavior', *Journal of Management Education*, 31(30), pp. 321–338.

Rungtusanatham, Salvador, F. (2008) 'From Mass Production to Mass Customization: Hindrance factors, structural inertia and transition hazards', *Production and Operations Management*, 17, pp. 385–396.

Salvato, C., Rerup, C. (2011) 'Beyond collective entities; multilevel research on organizational routines and capabilities', *Journal of Management*, 37(2), pp. 468–490.

Sarkis, J. (2001) 'Benchmarking for agility', *Benchmarking: An International Journal*, 8(2), pp. 88–107.

Schein, E. (1985) *Organizational Culture and Leadership: A Dynamic View*. Jossey-Bass, San Francisco, CA.

Senge, P. (1990) *The Fifth Discipline*. New York, NY: Doubleday.

Sharp, J. M., Irani, Z., Desai, S. (1999) 'Working towards agile manufacturing in UK industry', *International Journal of Production Economics* 62(1), pp. 155–169.

Sheather, G. and Hanna, D. (2000) 'Towards an integrated supply network model', *The Journal of Enterprise Research Management, Australasian Production and Inventory Control Society*, 3(3), pp. 5–10.

Sohal, A. S. (1996) 'Developing a lean production organization: an Australian case study', International Journal of Operations & Production Management, 16(2), pp. 91–102.

Stalk, G. (1998) 'The time—the next source of competitive advantage', *Harvard Business Review*, Jul–Aug 1988, pp. 14–51.

Strauss, A. and Corbin, J. (1990) *Basics of Qualitative Research: Grounded Theory Procedures and Techniques*. London: Sage.

Su, J. C. P., Chang, Y., Ferguson, M., and Ho, J. C. (2010) 'The impact of delayed differentiation in make-to-order environments', *International Journal of Production Research*, 48(19), pp. 5809–5829.

Ohno, T. (1998) *Toyota Production System Beyond Large Scale Production*. Productivity Press, New York, NY.

Tang, Z., Chen, X., and Xiao, J. (2005b) 'Operational tactics and tenets of a new manufacturing paradigm 'instant customization', *International Journal of Production Research*, 43(14), pp. 2873–2894.

Tang, Z., Chen, X., and Xiao, J. (2010) 'Using the classic ground theory approach to understand consumer purchase decision in relation to the first customized products', *Journal of Product and Brand Management*, 19(3), pp. 181–197.

Tersine, R. J., Wacker, J. G. (2000) 'Customer Aligned Inventory Strategies: Agility Maxims', *International Journal of Agile Management System*, 2(2), pp. 114–120.

Thaler, R. (1980) 'Towards a Positive Theory of Consumer Choice', *Journal of Economic Behavior and Organization*, 1(1).

Treacy, M., and Wiersema, F. (1993) 'Customer intimacy and other value disciplines', *Harvard Business Review*.

Tu, Q., Vonderembse, M. A., and Ragu-Nathan, T. S. (2001) 'The impact of time-based manufacturing practices on mass customization and value to customer', *Journal of Operations Management*, 19, pp. 201–17.

Ulrich, K. T. (1995) 'The role of product architecture in the manufacturing firm', Research Policy, 24(3), pp. 419–40.

Varadarajan, P. R., Jayachandran, S., Gimeno, J. (1999) *The Theory of Multimarket Competition: A Synthesis and Implications for Market Strategy*, 63(3), pp. 49–66.

Victor, B., and Boynton, A. (1998) *Invented Here: Maximizing Your Organization's Internal Growth and Profitability*. Boston, MA: Harvard Business School Press.

Vonodh, S., Sundararai, G., Devadasan, S. R., Kuttalingam, D., Rajanavagam, D. (2010) 'Amalgamation of mass customization and agile manufacturing concepts; the theory and implementation study in an electronics switches manufacturing company', *International Journal of Production Research*, 48(7), pp. 2141–2164.

Venkatraman, V. (2011) *Inspired Take Your Dream from Concept to Shelf.* John Wiley & Sons, NJ.

Veal, A. J. (2005) *Business Research Methods: A Managerial Approach*, 2nd Ed., Pearson Education Australia.

Wang, Z,.Chen, L., Zhao, X., Zhou, W. (2014) 'Modularity in building mass customization capability: The mediating effects of customization knowledge utilization and business process improvement,' Science Direct, 34(11), pp. 678–687.

Walton, R. E. (1986) 'From control to commitment in the workplace', *Harvard Business Review*, 63, pp. 76–84.

Weick, K. E. (1979) *The Social Psychology of Organizing*, 2nd Ed. Reading, MA: Addison-Wesley.

Wickens, P. (1993) 'Lean production and beyond the system, its critics and the future', *Human Resource Management Journal*, 3(4), pp. 75–89.

Wilkins, A. and Dyer, W. Jr. (1988) 'Toward culturally sensitive theories of cultural change', *Academy of Management Review*, 13, pp. 522–33.

Wiseman, Liz (2017) *Multipliers: How the Best Leaders Make Everyone Smarter.* Harper Collins Publishers. New York, NY.

White, B. J. (1988) 'Accelerating quality improvement'. Presentation to the conference board, *Total Quality Performance Conference*, New York, NY.

White, G. P. (1996) 'A meta-analysis model of manufacturing capabilities', *Journal of Operations Management*, 14, pp. 315–331.

Wind, J., and Rangaswamy, A. (2001) 'Mass customization: The next revolution in mass customization', *Journal of Interactive Marketing*, 15(1), pp. 13–32.

Womack, J. P., Jones, D. T., and Roos, D. (1991) *The Machine That Changed the World: The Story of Lean Production*. New York, NY: Macmillan.

Yilmaz, C., and Ergun, E. (2008) 'Organizational culture and firm effectiveness: An examination of relative effects of culture traits and the balanced culture hypothesis in an emerging economy', *Journal of World Business*, 43, pp. 290–306.

Yin, R. K. (1994) Case Study Research: Design and Methods. Beverly Hills, CA: Sage.

Yin, R. K. (1991) Applications of case study research. Washington, DC: Cosmo Corp. p. 52.

Yinan, Q., Tang, M., Zhang, M. (2014) 'Mass Customization in Flat Organization', *Journal of Applied Research and Technology*, 2014,12(2).

Zhang, Z. and Sharifi, H. (2000) 'A methodology for achieving agility in manufacturing organizations', *International Journal of Operations and Production Management*, 20(4), pp. 496–512.

Zipkin, P. (2001) 'The limits of mass customization', *MIT Sloan Management Review*, 42, pp. 81–87.

Printed in the United States
by Baker & Taylor Publisher Services